Cram101 Textbook Outlines to accompany:

Multivariate Data Analysis

Hair and Anderson and Tatham and Black, 5th Edition

An Academic Internet Publishers (AIPI) publication (c) 2007.

You have a discounted membership at www.Cram101.com with this book.

Get all of the practice tests for the chapters of this textbook, and access in-depth reference material for writing essays and papers. Here is an example from a Cram101 Biology text:

When you need problem solving help with math, stats, and other disciplines, www.Cram101.com will walk through the formulas and solutions step by step.

With Cram101.com online, you also have access to extensive reference material.

You will nail those essays and papers. Here is an example from a Cram101 Biology text:

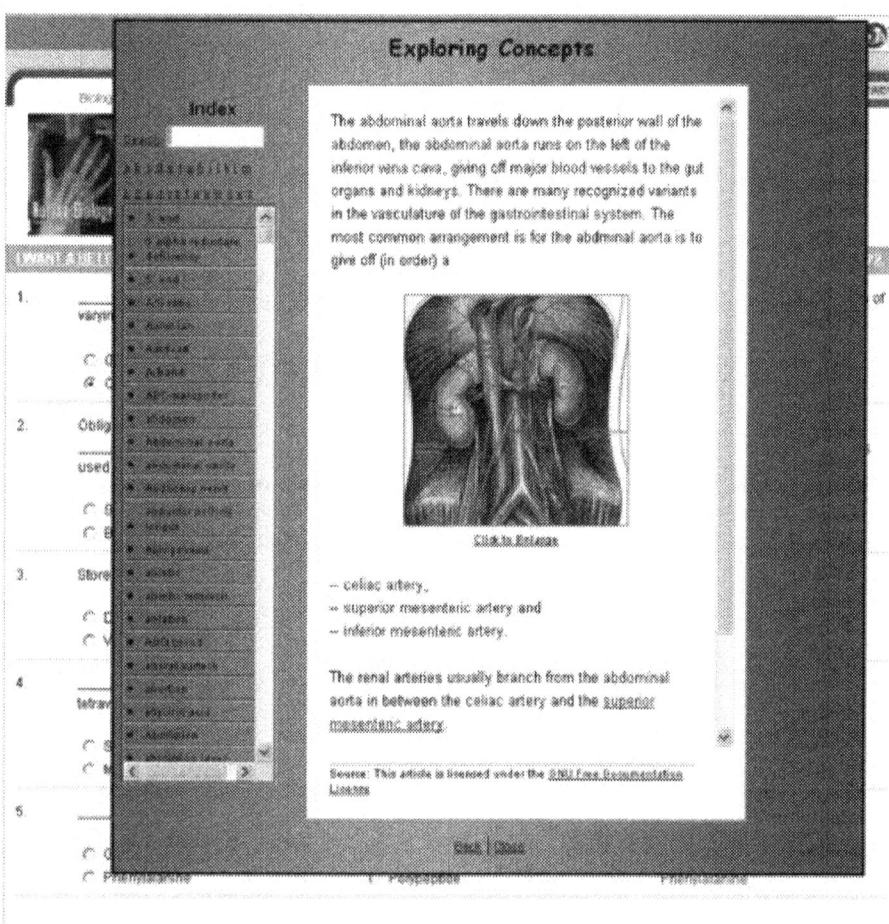

Visit **www.Cram101.com**, click Sign Up at the top of the screen, and enter DK73DW1119 in the promo code box on the registration screen. Access to www.Cram101.com is normally $9.95, but because you have purchased this book, your access fee is only $4.95. Sign up and stop highlighting textbooks forever.

Learning System

Cram101 Textbook Outlines is a learning system. The notes in this book are the highlights of your textbook, you will never have to highlight a book again.

How to use this book. Take this book to class, it is your notebook for the lecture. The notes and highlights on the left hand side of the pages follow the outline and order of the textbook. All you have to do is follow along while your intructor presents the lecture. Circle the items emphasized in class and add other important information on the right side. With Cram101 Textbook Outlines you'll spend less time writing and more time listening. Learning becomes more efficient.

Cram101.com Online

Increase your studying efficiency by using Cram101.com's practice tests and online reference material. It is the perfect complement to Cram101 Textbook Outlines. Use self-teaching matching tests or simulate in-class testing with comprehensive multiple choice tests, or simply use Cram's true and false tests for quick review. Cram101.com even allows you to enter your in-class notes for an integrated studying format combining the textbook notes with your class notes.

Visit **www.Cram101.com**, click Sign Up at the top of the screen, and enter **DK73DW1119** in the promo code box on the registration screen. Access to www.Cram101.com is normally $9.95, but because you have purchased this book, your access fee is only $4.95. Sign up and stop highlighting textbooks forever.

Multivariate Data Analysis
Hair and Anderson and Tatham and Black, 5th

CONTENTS

Multivariate analysis	Multivariate analysis in statistics describes a collection of procedures which involve observation and analysis of more than one statistical variable at a time.
Multivariate	In statistics, in multivariate data, each data point has more than one scalar component, and often one is concerned with correlations between the components.
Measurement	Measurement generally refers to the process of estimating or determining the ratio of a magnitude of a quantitative property or relation to a unit of the same type of quantitative property or relation.
Univariate	In statistics, in univariate data, each data point has only one scalar component. Or, when the statistical technique to be used contains only one dependent variable.
Independent variable	An independent variable is presumed to cause or determine a dependent variable. It can be changed as required and its values do not represent a problem requiring explanation in an analysis, but are taken simply as given.
Partial correlation	A partial correlation is a correlation between two variables when the effects of one or more related variables are removed.
Association	In statistics, an association comes from two variables that are related. Many people confuse association with cauzation. Association does not imply cauzation.
Correlation	Correlation indicates the strength and direction of a linear relationship between two random variables. In general statistical usage, correlation refers to the departure of two variables from independence.
Variable	A variable is a symbol denoting a quantity or symbolic representation. In mathematics, a variable often represents an unknown quantity; in computer science, it represents a place where a quantity can be stored.
Variance	The variance of a random variable is a measure of its statistical dispersion, indicating how far from the expected value its values typically are. The variance of a real-valued random variable is its second central moment, and it also happens to be its second cumulant.
Residual	Error is a misnomer; an error is the amount by which an observation differs from its expected value; the latter being based on the whole population from which the statistical unit was chosen randomly. A residual, on the other hand, is an observable estimate of the unobservable error.
Sets	Sets are collections of objects considered as a whole. The objects of sets are called elements or members. The elements of a set can be anything: numbers, people, letters of the alphabet, other sets, and so on. Sets are conventionally denoted with capital letters, A, B, C, etc. Two sets A and B are said to be equal, written A = B, if they have the same members.
Dependent variable	A variable that changes as a function of a change to another variable is called a dependent variable.
Regression analysis	Regression analysis is any statistical method where the mean of one or more random variables is predicted conditioned on other (measured) random variables.
Dummy variable	A dummy variable is a notation for a place or places in an expression, into which some definite substitution may take place, or with respect to which some operation (summation or quantification, to give two examples) may take place. The idea is related to, but somewhat deeper and more complex than, that of a placeholder (a symbol that will later be replaced by some literal string), or a wildcard character that stands for an unspecified symbol.
Effect size	An effect size describes how large the relationship is between two variables. This information is important in scientific research. Often it is useful to know not only whether an experiment had an effect, but also the size of any effects.
Population	A population is a set of entities concerning which statistical inferences are to be drawn, often based on a random sample taken from the population.

Mean	For a real-valued random variable X, the mean is the expectation of X. If the expectation does not exist, then the random variable has no mean. For a data set, the mean is just the sum of all the observations divided by the number of observations.
Multicolline-rity	Multicollinearity refers to linear inter-correlation among variables. Simply put, if nominally "different" measures actually quantify the same phenomenon to a significant degree -- i.e., wherein the variables are accorded different names and perhaps employ different numeric measurement scales but correlate highly with each other -- they are redundant.
Power	The power of a statistical test is the probability that the test will reject a false null hypothesis, or in other words that it will not make a Type II error. The higher the power, the greater the chance of obtaining a statistically significant result when the null hypothesis is false.
Probability	Probability is the ratio of the number of favorable outcomes to the number of possible outcomes.
Hypothesis	A hypothesis is a proposed explanation for a phenomenon. A scientific hypothesis must be testable and based on previous observations or extensions of scientific theories.
Null hypothesis	A null hypothesis, H_0, is a hypothesis set up to be nullified or refuted in order to support an alternative hypothesis. When used, the null hypothesis is presumed true until statistical evidence in the form of a hypothesis test indicates otherwise.
Statistical significance	A result is has statistical significance if it is unlikely to have occurred by chance, given that a presumed null hypothesis is true.
Type I error	A false positive, also called a Type I error, exists when a test incorrectly reports that it has found a result where none really exists.
Sample size	Sensitivity can be increased by using statistical controls, by increasing the reliability of measures (as in psychometric reliability), and by increasing the size of the sample. Increasing sample size is the most commonly used method for increasing statistical power.
Sample	A sample is that part of a population which is actually observed. In normal scientific practice, we demand that it is selected in such a way as to avoid presenting a biased view of the population.
Significance level	The significance level of a test is the maximum probability of accidentally rejecting a true null hypothesis (a decision known as a Type I error). The significance of a result is also called its p-value; the smaller the p-value, the more significant the result is said to be.
Reliability	Reliability has been defined in different ways by different authors. Perhaps the best way to look at reliability is the extent to which the measurements resulting from a test are the result of characteristics of those being measured.
Type II error	In statistics, a false negative, also called a Type II error or miss, exists when a test incorrectly reports that a result was not detected, when it was really present.
Analysis of variance	Analysis of variance (ANOVA) is a collection of statistical models and their associated procedures which compare means by splitting the overall observed variance into different parts.
Statistical Inference	Statistical inference is inference about a population from a random sample drawn from it or, more generally, about a random process from its observed behavior during a finite period of time. It includes: point estimation, interval estimation, hypothesis testing (or statistical significance testing) prediction
Statistic	A statistic is a characteristic of a sample drawn from a population.
Multidimensi-nal scaling	Multidimensional scaling is a set of related statistical techniques often used in data visualization for exploring similarities or dissimilarities in data. An algorithm starts with a matrix of item-item similarities, then assigns a location of each item in a low-dimensional space, suitable for graphing or 3D visualization.

Go to **Cram101.com** for the Practice Tests for this Chapter.

Go to **Cram101.com** for the Practice Tests for this Chapter.
And, **NEVER** highlight a book again!

Sampling	Sampling is that part of statistical practice concerned with the selection of individual observations intended to yield some knowledge about a population of concern, especially for the purposes of statistical inference.
Distribution	A distribution is a list of the values that a variable takes in a sample. It is usually a list, ordered by quantity.
Predictor variable	The predictor variable is manipulated by the experimenter. By attempting to isolate all other factors, one can determine the influence of the independent variable on the dependent variable.
Multiple regression	A multiple regression is a linear regression with more than one covariate (predictor variable). It can be viewed as a simple case of canonical correlation. An equation used to predict a dependent variable, y from two independents, u and v is: $y = \beta_0 + \beta_1 u + \beta_2 v + \beta_3 u2 + \beta_4 uv + \beta_5 v2$
Nominal	In nominal measurement numerals are assigned to objects as labels or names. If two entities have the same number associated with them, they belong to the same category, and that is the only significance that they have. The only comparisons that can be made between variable values are equality and inequality.
Ordinal	In ordinal measurement the numbers assigned to objects represent the rank order (1st, 2nd, 3rd etc) of the entities measured. Comparisons of greater and less can be made, in addition to equality and inequality. However operations such as conventional addition and subtraction are still without meaning.
Level of measurement	The level of measurement of a variable in mathematics and statistics is a classification that was proposed in order to describe the nature of information contained within numbers assigned to objects and, therefore, within the variable.
Interval scale	I an interval scale the numbers assigned to objects have all the features of ordinal measurements. In addition, equal differences between measurements represent equivalent intervals; i.e. differences between arbitrary pairs of measurements can be meaningfully compared.
Statistical power	Statistical power depends on the significance criterion, the size of the difference or the strength of the similarity (that is, the effect size) in the population, and the sensitivity of the data.
Alpha level	Among all the sets of possible values, we must choose one that we think represents the most extreme evidence against the hypothesis. That is called the critical region of the test statistic. The probability of the test statistic falling in the critical region when the hypothesis is correct is called the alpha level for the test.
Deviation	A deviation is the difference between an observed value and the expected value of a variable or function.
Standard deviation	The standard deviation is the most commonly used measure of statistical dispersion. Simply put, it measures how spread out the values in a data set are. The standard deviation is defined as the square root of the variance.
Range	In descriptive statistics, the range is the length of the smallest interval which contains all the data. It is calculated by subtracting the smallest observations from the greatest and provides an indication of statistical dispersion.
Discriminant analysis	Linear discriminant analysis and the related Fisher's linear discriminant are used in machine learning to find the linear combination of features which best separate two or more classes of object or event.
Discriminant	A discriminant is an expression that discriminates qualities of algebraic structures. The concept applies to polynomials, conic sections, quadratic forms, and algebraic number fields.
Multivariate analysis of variance	Multivariate analysis of variance is an extension of analysis of variance (ANOVA) methods to cover cases where there is more than one dependent variable and where the dependent variables cannot simply be combined.
Covariance	Intuitively, covariance is the measure of how much two variables vary together. That is to say, the

covariance becomes more positive for each pair of values which differ from their mean in the same direction, and becomes more negative with each pair of values which differ from their mean in opposite directions.

Analysis of covariance	Analysis of covariance is an old-fashioned name for a linear regression model with one continuous explanatory variable and one or more factors. The name exists for historical reasons, but there is no particular reason to distinguish the method from the general purpose linear model.
Combinations	Combinations are un-ordered collections of unique elements. The order of the elements is not important
Canonical correlation	Canonical correlation analysis seeks vectors a and b such that the random variables a'X and b'Y maximize the correlation p = cor(a'X, b'Y)..
Regression equation	The regression equation represents the relation between selected values of one variable (x) and observed values of the other (y); it permits the prediction of the most probable values of y.
Data reduction	All of statistical analysis provides data reduction, taking large amounts of information and reducing it into a more understandable form. Data reduction in this sense always results in a loss of information.
Inferential statistics	Inferential statistics or statistical induction comprises the use of statistics to make inferences concerning some unknown aspect (usually a parameter) of a population.
Bootstrapping	Bootstrapping is a statistical method for estimating the sampling distribution of an estimator by sampling with replacement from the original sample, most often with the purpose of deriving robust estimates of standard errors and confidence intervals of a population parameter like a mean, median, proportion, odds ratio, correlation coefficient or regression coefficient.
Exhaustive	In set theory, exhaustive is an attribute of rule(s) for defining the inclusion/exclusion of elements of a collection of sets such that no non-exclusive elements remain which cannot definitely be included in one and only one set.
Outlier	In statistics, an outlier is a single observation "far away" from the rest of the data. In most samplings of data, some data points will be further away from their expected values than what is deemed reasonable. This can be due to systematic error or faults in the theory that generated the expected values.
Applied research	Applied research is done to solve specific, practical questions; its primary aim is not to gain knowledge for its own sake. It can be exploratory, but is usually descriptive. It is almost always done on the basis of basic research.
Generalizability	Generalizability describes whether the results of individual studies and investigation samples can be applied to other studies and investigation samples.

Multivariate	In statistics, in multivariate data, each data point has more than one scalar component, and often one is concerned with correlations between the components.
Univariate	In statistics, in univariate data, each data point has only one scalar component. Or, when the statistical technique to be used contains only one dependent variable.
Outlier	In statistics, an outlier is a single observation "far away" from the rest of the data. In most samplings of data, some data points will be further away from their expected values than what is deemed reasonable. This can be due to systematic error or faults in the theory that generated the expected values.
Variable	A variable is a symbol denoting a quantity or symbolic representation. In mathematics, a variable often represents an unknown quantity; in computer science, it represents a place where a quantity can be stored.
Multivariate analysis	Multivariate analysis in statistics describes a collection of procedures which involve observation and analysis of more than one statistical variable at a time.
Sample	A sample is that part of a population which is actually observed. In normal scientific practice, we demand that it is selected in such a way as to avoid presenting a biased view of the population.
Dummy variable	A dummy variable is a notation for a place or places in an expression, into which some definite substitution may take place, or with respect to which some operation (summation or quantification, to give two examples) may take place. The idea is related to, but somewhat deeper and more complex than, that of a placeholder (a symbol that will later be replaced by some literal string), or a wildcard character that stands for an unspecified symbol.
Distribution	A distribution is a list of the values that a variable takes in a sample. It is usually a list, ordered by quantity.
Independent variable	An independent variable is presumed to cause or determine a dependent variable. It can be changed as required and its values do not represent a problem requiring explanation in an analysis, but are taken simply as given.
Mean	For a real-valued random variable X, the mean is the expectation of X. If the expectation does not exist, then the random variable has no mean. For a data set, the mean is just the sum of all the observations divided by the number of observations.
Histogram	A histogram is a graphical display of tabulated frequencies. That is, a histogram is the graphical version of a table which shows what proportion of cases fall into each of several or many specified categories. The categories are usually specified as nonoverlapping intervals of some variable. The categories (bars) must be adjacent.
Frequency	Frequency is the measurement of the number of times that a repeated event occurs per unit of time.
Normal distribution	The normal distribution is an extremely important probability distribution in many fields. It is a family of distributions of the same general form, differing in their location and scale parameters: the mean and standard deviation. The standard normal distribution is the normal distribution with a mean of zero and a standard deviation of one
Variance	The variance of a random variable is a measure of its statistical dispersion, indicating how far from the expected value its values typically are. The variance of a real-valued random variable is its second central moment, and it also happens to be its second cumulant.
Range	In descriptive statistics, the range is the length of the smallest interval which contains all the data. It is calculated by subtracting the smallest observations from the greatest and provides an indication of statistical dispersion.

Predictor variable	The predictor variable is manipulated by the experimenter. By attempting to isolate all other factors, one can determine the influence of the independent variable on the dependent variable.
Population	A population is a set of entities concerning which statistical inferences are to be drawn, often based on a random sample taken from the population.
Linear regression	In statistics, linear regression is a method of estimating the conditional expected value of one variable y given the values of some other variable or variables x. The variable of interest, y, is conventionally called the "response variable".
Residual	Error is a misnomer; an error is the amount by which an observation differs from its expected value; the latter being based on the whole population from which the statistical unit was chosen randomly. A residual, on the other hand, is an observable estimate of the unobservable error.
Kurtosis	Kurtosis is a measure of the "peakedness" of the distribution of a real-valued random variable. Higher kurtosis means more of the variance is due to infrequent extreme deviations, as opposed to frequent modestly-sized deviations.
Slope	The slope is commonly used to describe the measurement of the steepness, incline or grade of a straight line. A higher slope value indicates a steeper incline.
Random sample	A sample is a subset chosen from a population for investigation. A random sample is one chosen by a method involving an unpredictable component, in the sense that the selection of any element of the population is independent of the selection of any other element.
Bias	A bias is a prejudice in a general or specific sense, usually in the sense for having a predilection to one particular point of view or ideological perspective. However, one is generally only said to be biased if one's powers of judgement are influenced by the biases one holds, to the extent that one's views could not be taken as being neutral or objective, but instead as subjective.
Probability distribution	Every random variable gives rise to a probability distribution, containing the most important information about the variable. If X is a random variable, the corresponding probability distribution assigns to the interval (a, b) the probability $Pr(a \leq X \leq b)$, i.e. the probability that the variable X will take a value in the interval (a, b).
Probability	Probability is the ratio of the number of favorable outcomes to the number of possible outcomes.
Unimodal	A function f(x) between two ordered sets is unimodal if for some value m (the mode), it is monotonically increasing for $x \leq m$ and monotonically decreasing for $x \geq m$. In that case, the maximum value of f(x) is f(m).
Deviation	A deviation is the difference between an observed value and the expected value of a variable or function.
Dependent variable	A variable that changes as a function of a change to another variable is called a dependent variable.
Scatterplot	A scatterplot is used in statistics to visually display and compare two or more sets of related quantitative, or numerical, data by displaying only finitely many points, each having a coordinate on a horizontal and a vertical axis.
Skewness	Skewness is a measure of the asymmetry of the distribution of a real-valued random variable. Skewness, the third standardized moment, is written as γ1 and defined as $\gamma 1 = \mu^3 / \sigma^3$ where μ^3 is the third moment about the mean and σ is the standard deviation.
Negatively	Roughly speaking, a distribution is negatively skewed if the lower tail is longer than the

skewed	higher tail.
Positively skewed	A distribution is positively skewed when is has a tail extending out to the right. When a distribution is positively skewed, the mean is greater than the median reflecting the fact that the mean is sensitive to each score in the distribution and is subject to large shifts when the sample is small and contains extreme scores.
Power	The power of a statistical test is the probability that the test will reject a false null hypothesis, or in other words that it will not make a Type II error. The higher the power, the greater the chance of obtaining a statistically significant result when the null hypothesis is false.
Sets	Sets are collections of objects considered as a whole. The objects of sets are called elements or members. The elements of a set can be anything: numbers, people, letters of the alphabet, other sets, and so on. Sets are conventionally denoted with capital letters, A, B, C, etc. Two sets A and B are said to be equal, written A = B, if they have the same members.
Statistical power	Statistical power depends on the significance criterion, the size of the difference or the strength of the similarity (that is, the effect size) in the population, and the sensitivity of the data.
Construct	A construct is a mathematical or conceptual model.
Stem and Leaf Plot	A stem and leaf plot is a graphical display of quantitative data that is similar to a histogram and is useful in visualizing the shape of a distribution. They contain more information than do histograms because, unlike in a histogram where bars are used, the individual data values are displayed in a table-like format, in order of increasing magnitude.
Expected value	The expected value (or mathematical expectation) of a random variable is the sum of the probability of each possible outcome of the experiment multiplied by its payoff ("value").
Linear relationship	A situation in which the best-fitting regression line is a straight line is called a linear relationship.
Correlation	Correlation indicates the strength and direction of a linear relationship between two random variables. In general statistical usage, correlation refers to the departure of two variables from independence.
Combinations	Combinations are un-ordered collections of unique elements. The order of the elements is not important
Multivariate analysis of variance	Multivariate analysis of variance is an extension of analysis of variance (ANOVA) methods to cover cases where there is more than one dependent variable and where the dependent variables cannot simply be combined.
Discriminant analysis	Linear discriminant analysis and the related Fisher's linear discriminant are used in machine learning to find the linear combination of features which best separate two or more classes of object or event.
Analysis of variance	Analysis of variance (ANOVA) is a collection of statistical models and their associated procedures which compare means by splitting the overall observed variance into different parts.
Discriminant	A discriminant is an expression that discriminates qualities of algebraic structures. The concept applies to polynomials, conic sections, quadratic forms, and algebraic number fields.
Statistical significance	A result is has statistical significance if it is unlikely to have occurred by chance, given that a presumed null hypothesis is true.
Quartile	In descriptive statistics, a quartile is any of the three values which divide the sorted data

set into four equal parts, so that each part represents 1/4th of the sample or population.

Median	The median is a number that separates the higher half of a sample, a population, or a probability distribution from the lower half. It is the middle value in a distribution, above and below which lie an equal number of values.
Complement	Generally a complement of X is something that together with X makes a complete whole; that supplies what X lacks.
Sample size	Sensitivity can be increased by using statistical controls, by increasing the reliability of measures (as in psychometric reliability), and by increasing the size of the sample. Increasing sample size is the most commonly used method for increasing statistical power.
Sampling	Sampling is that part of statistical practice concerned with the selection of individual observations intended to yield some knowledge about a population of concern, especially for the purposes of statistical inference.
Sampling error	When analyzing collected data, the samples observed differ in such things as means and standard deviations from the population from which the sample is taken. This is sampling error and is controlled by ensuring that, as much as possible, the samples taken have no systematic characteristics and are a true random sample from all possible samples.
Association	In statistics, an association comes from two variables that are related. Many people confuse association with cauzation. Association does not imply cauzation.
Subset	A is a subset of a set B, if A is "contained" inside B. The relationship of one set being a subset of another is called inclusion. Every set is a subset of itself.
Estimator	An estimator is a function of the known sample data that is used to estimate an unknown population parameter; an estimate is the result from the actual application of the function to a particular set of data. Many different estimators are possible for any given parameter.
Standard deviation	The standard deviation is the most commonly used measure of statistical dispersion. Simply put, it measures how spread out the values in a data set are. The standard deviation is defined as the square root of the variance.
Covariance	Intuitively, covariance is the measure of how much two variables vary together. That is to say, the covariance becomes more positive for each pair of values which differ from their mean in the same direction, and becomes more negative with each pair of values which differ from their mean in opposite directions.
Generalizability	Generalizability describes whether the results of individual studies and investigation samples can be applied to other studies and investigation samples.
Dichotomous variable	A variable with only two possible values is a dichotomous variable.
Test for randomness	The chi-square test is the most commonly used test for randomness of data, and is extremely sensitive to errors in pseudorandom sequence generators.
Standard score	A standard score (z) is a dimensionless quantity derived by subtracting the population mean from an individual (raw) score and then dividing the difference by the population standard deviation: $z = (X-\mu) / \sigma$.
Confidence Interval	A confidence interval is an interval between two numbers, where there is a certain specified level of confidence that a population parameter lies.
Multiple regression	A multiple regression is a linear regression with more than one covariate (predictor variable). It can be viewed as a simple case of canonical correlation. An equation used to predict a dependent variable, y from two independents, u and v is: $y = \beta_0 + \beta_1 u + \beta_2 v + \beta_3 u2 + \beta_4 uv + \beta_5 v2$

Statistic	A statistic is a characteristic of a sample drawn from a population.
Descriptive statistics	Descriptive statistics is a branch of statistics that denotes any of the many techniques used to summarize a set of data. In a sense, we are using the data on members of a set to describe the set.
Uniform Distribution	In mathematics, the uniform distributions are simple probability distributions. There are two types: the discrete uniform distribution; the continuous uniform distribution.
Significance level	The significance level of a test is the maximum probability of accidentally rejecting a true null hypothesis (a decision known as a Type I error). The significance of a result is also called its p-value; the smaller the p-value, the more significant the result is said to be.
Critical value	A critical value is the value corresponding to a given significance level. This cutoff value determines the boundary between those samples resulting in a test statistic that leads to rejecting the null hypothesis and those lead to a decision not to reject the null hypothesis.
Hypothesis	A hypothesis is a proposed explanation for a phenomenon. A scientific hypothesis must be testable and based on previous observations or extensions of scientific theories.
Hypothesis test	One may be faced with the problem of making a definite decision with respect to an uncertain hypothesis which is known only through its observable consequences. A statistical hypothesis test, or more briefly, hypothesis test, is an algorithm to state the alternative which minimizes certain risks.
Dispersion	In descriptive statistics, statistical dispersion is quantifiable variation of measurements of differing members of a population within the scale on which they are measured.
Regression analysis	Regression analysis is any statistical method where the mean of one or more random variables is predicted conditioned on other (measured) random variables.
Categorical variable	In a categorical variable, numerals are assigned to objects as labels or names. If two entities have the same number associated with them, they belong to the same category, and that is the only significance that they have. The only comparisons that can be made between variable values are equality and inequality.
Measurement	Measurement generally refers to the process of estimating or determining the ratio of a magnitude of a quantitative property or relation to a unit of the same type of quantitative property or relation.
Multivariate statistical analysis	Multivariate statistical analysis in statistics describes a collection of procedures which involve observation and analysis of more than one statistical variable at a time.
Statistical analysis	Statistical analysis refers to the branch of mathematics that deals with the collection, analysis, interpretation and presentation of masses of numerical data.

Multivariate	In statistics, in multivariate data, each data point has more than one scalar component, and often one is concerned with correlations between the components.
Variable	A variable is a symbol denoting a quantity or symbolic representation. In mathematics, a variable often represents an unknown quantity; in computer science, it represents a place where a quantity can be stored.
Partial correlation	A partial correlation is a correlation between two variables when the effects of one or more related variables are removed.
Correlation	Correlation indicates the strength and direction of a linear relationship between two random variables. In general statistical usage, correlation refers to the departure of two variables from independence.
Sampling	Sampling is that part of statistical practice concerned with the selection of individual observations intended to yield some knowledge about a population of concern, especially for the purposes of statistical inference.
Variance	The variance of a random variable is a measure of its statistical dispersion, indicating how far from the expected value its values typically are. The variance of a real-valued random variable is its second central moment, and it also happens to be its second cumulant.
Content validity	In psychometrics, content validity refers to the extent to which a measure represents all facets of a given social concept. For example, a depression scale may lack content validity if it only assesses the affective dimension of depression but fails to take into account the behavioral dimension.
Lower limit	The lower limit in a distribution is the smallest value.
Reliability	Reliability has been defined in different ways by different authors. Perhaps the best way to look at reliability is the extent to which the measurements resulting from a test are the result of characteristics of those being measured.
Range	In descriptive statistics, the range is the length of the smallest interval which contains all the data. It is calculated by subtracting the smallest observations from the greatest and provides an indication of statistical dispersion.
Dummy variable	A dummy variable is a notation for a place or places in an expression, into which some definite substitution may take place, or with respect to which some operation (summation or quantification, to give two examples) may take place. The idea is related to, but somewhat deeper and more complex than, that of a placeholder (a symbol that will later be replaced by some literal string), or a wildcard character that stands for an unspecified symbol.
Orthogonal	In mathematics, orthogonal is synonymous with perpendicular when used as a simple adjective that is not part of any longer phrase with a standard definition. It means at right angles.
Deviation	A deviation is the difference between an observed value and the expected value of a variable or function.
Mean	For a real-valued random variable X, the mean is the expectation of X. If the expectation does not exist, then the random variable has no mean. For a data set, the mean is just the sum of all the observations divided by the number of observations.
Standard deviation	The standard deviation is the most commonly used measure of statistical dispersion. Simply put, it measures how spread out the values in a data set are. The standard deviation is defined as the square root of the variance.
Measurement	Measurement generally refers to the process of estimating or determining the ratio of a magnitude of a quantitative property or relation to a unit of the same type of quantitative property or relation.

Data reduction	All of statistical analysis provides data reduction, taking large amounts of information and reducing it into a more understandable form. Data reduction in this sense always results in a loss of information.
Dependent variable	A variable that changes as a function of a change to another variable is called a dependent variable.
Elements	A set is a collection of objects considered as a whole. The objects of a set are called elements or members. The elements of a set can be anything: numbers, people, letters of the alphabet, other sets, and so on.
Sample	A sample is that part of a population which is actually observed. In normal scientific practice, we demand that it is selected in such a way as to avoid presenting a biased view of the population.
Multivariate analysis	Multivariate analysis in statistics describes a collection of procedures which involve observation and analysis of more than one statistical variable at a time.
Subset	A is a subset of a set B, if A is "contained" inside B. The relationship of one set being a subset of another is called inclusion. Every set is a subset of itself.
Power	The power of a statistical test is the probability that the test will reject a false null hypothesis, or in other words that it will not make a Type II error. The higher the power, the greater the chance of obtaining a statistically significant result when the null hypothesis is false.
Sets	Sets are collections of objects considered as a whole. The objects of sets are called elements or members. The elements of a set can be anything: numbers, people, letters of the alphabet, other sets, and so on. Sets are conventionally denoted with capital letters, A, B, C, etc. Two sets A and B are said to be equal, written A = B, if they have the same members.
Sample size	Sensitivity can be increased by using statistical controls, by increasing the reliability of measures (as in psychometric reliability), and by increasing the size of the sample. Increasing sample size is the most commonly used method for increasing statistical power.
Covariance	Intuitively, covariance is the measure of how much two variables vary together. That is to say, the covariance becomes more positive for each pair of values which differ from their mean in the same direction, and becomes more negative with each pair of values which differ from their mean in opposite directions.
Significance level	The significance level of a test is the maximum probability of accidentally rejecting a true null hypothesis (a decision known as a Type I error). The significance of a result is also called its p-value; the smaller the p-value, the more significant the result is said to be.
Generalizability	Generalizability describes whether the results of individual studies and investigation samples can be applied to other studies and investigation samples.
Mode	The mode is the value that has the largest number of observations, namely the most frequent value within a particular set of values.
Probability	Probability is the ratio of the number of favorable outcomes to the number of possible outcomes.
Principal components analysis	Principal components analysis is a technique for simplifying a dataset. It is a linear transformation that transforms the data to a new coordinate system such that the greatest variance by any projection of the data comes to lie on the first coordinate, the second greatest variance on the second coordinate, and so on.
Construct	A construct is a mathematical or conceptual model.
Combinations	Combinations are un-ordered collections of unique elements. The order of the elements is not

	important
A priori	In statistics, a priori knowledge refers to a knowledge of the actual population, rather than that estimated by observation.
Slope	The slope is commonly used to describe the measurement of the steepness, incline or grade of a straight line. A higher slope value indicates a steeper incline.
Origin	The point of intersection, where the axes meet, is called the origin normally labeled O. With the origin labeled O, we can name the x axis Ox and the y axis Oy. The x and y axes define a plane that can be referred to as the xy plane.
Association	In statistics, an association comes from two variables that are related. Many people confuse association with cauzation. Association does not imply cauzation.
Statistical significance	A result is has statistical significance if it is unlikely to have occurred by chance, given that a presumed null hypothesis is true.
Population	A population is a set of entities concerning which statistical inferences are to be drawn, often based on a random sample taken from the population.
Outlier	In statistics, an outlier is a single observation "far away" from the rest of the data. In most samplings of data, some data points will be further away from their expected values than what is deemed reasonable. This can be due to systematic error or faults in the theory that generated the expected values.
Multivariate analysis of variance	Multivariate analysis of variance is an extension of analysis of variance (ANOVA) methods to cover cases where there is more than one dependent variable and where the dependent variables cannot simply be combined.
Discriminant analysis	Linear discriminant analysis and the related Fisher's linear discriminant are used in machine learning to find the linear combination of features which best separate two or more classes of object or event.
Independent variable	An independent variable is presumed to cause or determine a dependent variable. It can be changed as required and its values do not represent a problem requiring explanation in an analysis, but are taken simply as given.
Analysis of variance	Analysis of variance (ANOVA) is a collection of statistical models and their associated procedures which compare means by splitting the overall observed variance into different parts.
Discriminant	A discriminant is an expression that discriminates qualities of algebraic structures. The concept applies to polynomials, conic sections, quadratic forms, and algebraic number fields.
Internal consistency	Internal consistency focuses on the degree to which the individual items are correlated with each other and is thus often called homogeneity. Several statistics fall within this category. The best known are Cronbach's alpha, the Kuder-Richardson Formula 20 (KR-20) and the Kuder-Richardson Formula 21 (KR-21).
Replication	Replication is repeating the creation of a phenomenon, so that the variability associated with the phenomenon can be estimated.
Sum of squares	Sum of squares is a concept that permeates much of inferential statistics and descriptive statistics. More properly, it is "the sum of the squared deviations". Mathematically, it is an unscaled, or unadjusted measure of dispersion. When scaled for the number of degrees of freedom, it becomes the variance, the sum of squares per degree of freedom.
Distribution	A distribution is a list of the values that a variable takes in a sample. It is usually a list, ordered by quantity.

Sampling error	When analyzing collected data, the samples observed differ in such things as means and standard deviations from the population from which the sample is taken. This is sampling error and is controlled by ensuring that, as much as possible, the samples taken have no systematic characteristics and are a true random sample from all possible samples.
Construct validity	In social science and psychometrics, construct validity refers to whether a scale measures the unobservable social construct (such as "fluid intelligence") that it purports to measure. The unobservable idea of a unidimensional easier-to-harder dimension must be "constructed" in the words of human language and graphics.
Statistic	A statistic is a characteristic of a sample drawn from a population.
Multiple regression	A multiple regression is a linear regression with more than one covariate (predictor variable). It can be viewed as a simple case of canonical correlation. An equation used to predict a dependent variable, y from two independents, u and v is: $y = \beta_0 + \beta_1 u + \beta_2 v + \beta_3 u2 + \beta_4 uv + \beta_5 v2$
Regression analysis	Regression analysis is any statistical method where the mean of one or more random variables is predicted conditioned on other (measured) random variables.
Canonical correlation	Canonical correlation analysis seeks vectors a and b such that the random variables a'X and b'Y maximize the correlation p = cor(a'X, b'Y)..

Least Squares	Least squares is a mathematical optimization technique which, when given a series of measured data, attempts to find a function which closely approximates the data (a "best fit"). It attempts to minimize the sum of the squares of the ordinate differences (called residuals) between points generated by the function and corresponding points in the data.
Regression analysis	Regression analysis is any statistical method where the mean of one or more random variables is predicted conditioned on other (measured) random variables.
Variable	A variable is a symbol denoting a quantity or symbolic representation. In mathematics, a variable often represents an unknown quantity; in computer science, it represents a place where a quantity can be stored.
Dummy variable	A dummy variable is a notation for a place or places in an expression, into which some definite substitution may take place, or with respect to which some operation (summation or quantification, to give two examples) may take place. The idea is related to, but somewhat deeper and more complex than, that of a placeholder (a symbol that will later be replaced by some literal string), or a wildcard character that stands for an unspecified symbol.
Range	In descriptive statistics, the range is the length of the smallest interval which contains all the data. It is calculated by subtracting the smallest observations from the greatest and provides an indication of statistical dispersion.
Independent variable	An independent variable is presumed to cause or determine a dependent variable. It can be changed as required and its values do not represent a problem requiring explanation in an analysis, but are taken simply as given.
Multiple regression	A multiple regression is a linear regression with more than one covariate (predictor variable). It can be viewed as a simple case of canonical correlation. An equation used to predict a dependent variable, y from two independents, u and v is: $y = \beta_0 + \beta_1 u + \beta_2 v + \beta_3 u2 + \beta_4 uv + \beta_5 v2$
Dependent variable	A variable that changes as a function of a change to another variable is called a dependent variable.
Coefficient of determination	The coefficient of determination R^2 is the proportion of a sample variance of a response variable that is "explained" by the predictor variables when a linear regression is done.
Regression equation	The regression equation represents the relation between selected values of one variable (x) and observed values of the other (y); it permits the prediction of the most probable values of y.
Sample size	Sensitivity can be increased by using statistical controls, by increasing the reliability of measures (as in psychometric reliability), and by increasing the size of the sample. Increasing sample size is the most commonly used method for increasing statistical power.
Sample	A sample is that part of a population which is actually observed. In normal scientific practice, we demand that it is selected in such a way as to avoid presenting a biased view of the population.
Power	The power of a statistical test is the probability that the test will reject a false null hypothesis, or in other words that it will not make a Type II error. The higher the power, the greater the chance of obtaining a statistically significant result when the null hypothesis is false.
Degrees of freedom	In fitting statistical models to data, the vectors of residuals are often constrained to lie in a space of smaller dimension than the number of components in the vector. That smaller dimension is the number of degrees of freedom for error.
Combinations	Combinations are un-ordered collections of unique elements. The order of the elements is not important

Go to **Cram101.com** for the Practice Tests for this Chapter.

Standardized regression coefficient	The way to remove the effects of having different units for the two variables in a regression analysis is to normalize the two variables by dividing their values by their standard deviations, then fit the straight line. The resulting slope is known as a standardized regression coefficient.
Regression coefficient	The regression coefficient is the slope of the straight line that most closely relates two correlated variables.
Variance	The variance of a random variable is a measure of its statistical dispersion, indicating how far from the expected value its values typically are. The variance of a real-valued random variable is its second central moment, and it also happens to be its second cumulant.
Mean	For a real-valued random variable X, the mean is the expectation of X. If the expectation does not exist, then the random variable has no mean. For a data set, the mean is just the sum of all the observations divided by the number of observations.
Multicolline-rity	Multicollinearity refers to linear inter-correlation among variables. Simply put, if nominally "different" measures actually quantify the same phenomenon to a significant degree -- i.e., wherein the variables are accorded different names and perhaps employ different numeric measurement scales but correlate highly with each other -- they are redundant.
Association	In statistics, an association comes from two variables that are related. Many people confuse association with cauzation. Association does not imply cauzation.
Correlation	Correlation indicates the strength and direction of a linear relationship between two random variables. In general statistical usage, correlation refers to the departure of two variables from independence.
Criterion variable	The variable in a study that is expected to change as a result of alteration of the independent variable is the criterion variable.
Parameter	A parameter is a characteristic of a population.
Population	A population is a set of entities concerning which statistical inferences are to be drawn, often based on a random sample taken from the population.
Linear regression	In statistics, linear regression is a method of estimating the conditional expected value of one variable y given the values of some other variable or variables x. The variable of interest, y, is conventionally called the "response variable".
Outlier	In statistics, an outlier is a single observation "far away" from the rest of the data. In most samplings of data, some data points will be further away from their expected values than what is deemed reasonable. This can be due to systematic error or faults in the theory that generated the expected values.
Intercept	An intercept is the coordinate of the point at which a curve cuts an axis. For example, an x-intercept or a y-intercept.
Residual	Error is a misnomer; an error is the amount by which an observation differs from its expected value; the latter being based on the whole population from which the statistical unit was chosen randomly. A residual, on the other hand, is an observable estimate of the unobservable error.
Slope	The slope is commonly used to describe the measurement of the steepness, incline or grade of a straight line. A higher slope value indicates a steeper incline.
Measurement	Measurement generally refers to the process of estimating or determining the ratio of a magnitude of a quantitative property or relation to a unit of the same type of quantitative property or relation.
Normal	The normal distribution is an extremely important probability distribution in many fields. It

distribution	is a family of distributions of the same general form, differing in their location and scale parameters: the mean and standard deviation. The standard normal distribution is the normal distribution with a mean of zero and a standard deviation of one
Distribution	A distribution is a list of the values that a variable takes in a sample. It is usually a list, ordered by quantity.
Probability	Probability is the ratio of the number of favorable outcomes to the number of possible outcomes.
Deviation	A deviation is the difference between an observed value and the expected value of a variable or function.
Partial correlation	A partial correlation is a correlation between two variables when the effects of one or more related variables are removed.
Scatterplot	A scatterplot is used in statistics to visually display and compare two or more sets of related quantitative, or numerical, data by displaying only finitely many points, each having a coordinate on a horizontal and a vertical axis.
Complement	Generally a complement of X is something that together with X makes a complete whole; that supplies what X lacks.
Significance level	The significance level of a test is the maximum probability of accidentally rejecting a true null hypothesis (a decision known as a Type I error). The significance of a result is also called its p-value; the smaller the p-value, the more significant the result is said to be.
Predictor variable	The predictor variable is manipulated by the experimenter. By attempting to isolate all other factors, one can determine the influence of the independent variable on the dependent variable.
Statistic	A statistic is a characteristic of a sample drawn from a population.
Standard deviation	The standard deviation is the most commonly used measure of statistical dispersion. Simply put, it measures how spread out the values in a data set are. The standard deviation is defined as the square root of the variance.
Standard error	The standard error of a measurement, value or quantity is the standard deviation of the process by which it was generated, after adjusting for sample size.
Confidence Interval	A confidence interval is an interval between two numbers, where there is a certain specified level of confidence that a population parameter lies.
Total sum of squares	The total sum of squares is the sum of the squares of the difference of the independent variable and its grand mean. Furthermore, total sum of squares = explained sum of squares + residual sum of squares.
Sum of squares	Sum of squares is a concept that permeates much of inferential statistics and descriptive statistics. More properly, it is "the sum of the squared deviations". Mathematically, it is an unscaled, or unadjusted measure of dispersion. When scaled for the number of degrees of freedom, it becomes the variance, the sum of squares per degree of freedom.
Multivariate	In statistics, in multivariate data, each data point has more than one scalar component, and often one is concerned with correlations between the components.
Baseline	Baseline is a starting point or condition against which future changes are measured.
Point estimate	A point estimate is an estimate of a population parameter that is a single numerical value.
Confidence Level	The confidence level is the range (with a specified value of uncertainty, usually expressed in percent) within which the true value of a measured quantity exists.

Interaction	Interaction is a kind of action which occurs as two or more objects have an effect upon one another. The idea of a two-way effect is essential in the concept of interaction instead of a one-way causal effect.
Statistical power	Statistical power depends on the significance criterion, the size of the difference or the strength of the similarity (that is, the effect size) in the population, and the sensitivity of the data.
Generalizability	Generalizability describes whether the results of individual studies and investigation samples can be applied to other studies and investigation samples.
Sets	Sets are collections of objects considered as a whole. The objects of sets are called elements or members. The elements of a set can be anything: numbers, people, letters of the alphabet, other sets, and so on. Sets are conventionally denoted with capital letters, A, B, C, etc. Two sets A and B are said to be equal, written A = B, if they have the same members.
Univariate	In statistics, in univariate data, each data point has only one scalar component. Or, when the statistical technique to be used contains only one dependent variable.
Random Variable	A variable characterized by random behavior in assuming its different possible values is a random variable. Mathematically, it is described by its probability distribution, which specifies the possible values of a random variable together with the probability associated (in an appropriate sense) with each value.
Bias	A bias is a prejudice in a general or specific sense, usually in the sense for having a predilection to one particular point of view or ideological perspective. However, one is generally only said to be biased if one's powers of judgement are influenced by the biases one holds, to the extent that one's views could not be taken as being neutral or objective, but instead as subjective.
Levels of measurement	The levels of measurement of a variable in mathematics and statistics is a classification that was proposed in order to describe the nature of information contained within numbers assigned to objects and, therefore, within the variable.
Statistical significance	A result is has statistical significance if it is unlikely to have occurred by chance, given that a presumed null hypothesis is true.
Elements	A set is a collection of objects considered as a whole. The objects of a set are called elements or members. The elements of a set can be anything: numbers, people, letters of the alphabet, other sets, and so on.
Effect size	An effect size describes how large the relationship is between two variables. This information is important in scientific research. Often it is useful to know not only whether an experiment had an effect, but also the size of any effects.
Sampling	Sampling is that part of statistical practice concerned with the selection of individual observations intended to yield some knowledge about a population of concern, especially for the purposes of statistical inference.
Linear relationship	A situation in which the best-fitting regression line is a straight line is called a linear relationship.
Continuous variable	A quantitative variable with an infinite number of attributes is a Continuous variable. FOR EXAMPLE, distance or length.
Dispersion	In descriptive statistics, statistical dispersion is quantifiable variation of measurements of differing members of a population within the scale on which they are measured.
Histogram	A histogram is a graphical display of tabulated frequencies. That is, a histogram is the graphical version of a table which shows what proportion of cases fall into each of several

or many specified categories. The categories are usually specified as nonoverlapping intervals of some variable. The categories (bars) must be adjacent.

Midrange

The midrange of a set of statistical data values is the arithmetic mean of the smallest and largest values in the set. It is highly sensitive to outliers and ignores all but two data points; therefore it is rarely used in statistical analysis.

Time series

In statistics and signal processing, a time series is a sequence of data points, measured typically at successive times, spaced apart at uniform time intervals. Time series analysis comprises methods that attempt to understand such time series, often either to understand the underlying theory of the data points, or to make forecasts.

Direct correlation

A direct correlation is a positive correlation: a correlation in which large values of one variable are associated with large values of the other; the correlation coefficient is between 0 and -1.

Hypothesis

A hypothesis is a proposed explanation for a phenomenon. A scientific hypothesis must be testable and based on previous observations or extensions of scientific theories.

Test Statistic

A test statistic is a summary value (often a summary statistic) of a data set that is compared with a statistical distribution to determine whether the data set differs from that expected under a null hypothesis.

Sampling error

When analyzing collected data, the samples observed differ in such things as means and standard deviations from the population from which the sample is taken. This is sampling error and is controlled by ensuring that, as much as possible, the samples taken have no systematic characteristics and are a true random sample from all possible samples.

Simple effect

A simple effect of an independent variable is the effect at a single level of another variable. Often they are computed following a significant interaction.

Critical value

A critical value is the value corresponding to a given significance level. This cutoff value determines the boundary between those samples resulting in a test statistic that leads to rejecting the null hypothesis and those lead to a decision not to reject the null hypothesis.

Lower limit

The lower limit in a distribution is the smallest value.

Stepwise Regression

Stepwise regression is an incremental approach to multiple regression where the order of entry of predictor variables is determined systematically.

Multivariate statistical analysis

Multivariate statistical analysis in statistics describes a collection of procedures which involve observation and analysis of more than one statistical variable at a time.

Statistical analysis

Statistical analysis refers to the branch of mathematics that deals with the collection, analysis, interpretation and presentation of masses of numerical data.

Go to **Cram101.com** for the Practice Tests for this Chapter.
And, **NEVER** highlight a book again!

Discriminant analysis	Linear discriminant analysis and the related Fisher's linear discriminant are used in machine learning to find the linear combination of features which best separate two or more classes of object or event.
Discriminant	A discriminant is an expression that discriminates qualities of algebraic structures. The concept applies to polynomials, conic sections, quadratic forms, and algebraic number fields.
Independent variable	An independent variable is presumed to cause or determine a dependent variable. It can be changed as required and its values do not represent a problem requiring explanation in an analysis, but are taken simply as given.
Power	The power of a statistical test is the probability that the test will reject a false null hypothesis, or in other words that it will not make a Type II error. The higher the power, the greater the chance of obtaining a statistically significant result when the null hypothesis is false.
Variable	A variable is a symbol denoting a quantity or symbolic representation. In mathematics, a variable often represents an unknown quantity; in computer science, it represents a place where a quantity can be stored.
Multiple regression	A multiple regression is a linear regression with more than one covariate (predictor variable). It can be viewed as a simple case of canonical correlation. An equation used to predict a dependent variable, y from two independents, u and v is: $y = \beta_0 + \beta_1 u + \beta_2 v + \beta_3 u2 + \beta_4 uv + \beta_5 v2$
Intercept	An intercept is the coordinate of the point at which a curve cuts an axis. For example, an x-intercept or a y-intercept.
Regression analysis	Regression analysis is any statistical method where the mean of one or more random variables is predicted conditioned on other (measured) random variables.
Multivariate	In statistics, in multivariate data, each data point has more than one scalar component, and often one is concerned with correlations between the components.
Sample	A sample is that part of a population which is actually observed. In normal scientific practice, we demand that it is selected in such a way as to avoid presenting a biased view of the population.
Covariance	Intuitively, covariance is the measure of how much two variables vary together. That is to say, the covariance becomes more positive for each pair of values which differ from their mean in the same direction, and becomes more negative with each pair of values which differ from their mean in opposite directions.
Dependent variable	A variable that changes as a function of a change to another variable is called a dependent variable.
Statistical significance	A result is has statistical significance if it is unlikely to have occurred by chance, given that a presumed null hypothesis is true.
Categorical variable	In a categorical variable, numerals are assigned to objects as labels or names. If two entities have the same number associated with them, they belong to the same category, and that is the only significance that they have. The only comparisons that can be made between variable values are equality and inequality.
Mean	For a real-valued random variable X, the mean is the expectation of X. If the expectation does not exist, then the random variable has no mean. For a data set, the mean is just the sum of all the observations divided by the number of observations.
Z score	A z score is a dimensionless quantity derived by subtracting the population mean from an individual (raw) score and then dividing the difference by the population standard deviation:

Correlation	Correlation indicates the strength and direction of a linear relationship between two random variables. In general statistical usage, correlation refers to the departure of two variables from independence.
Measurement	Measurement generally refers to the process of estimating or determining the ratio of a magnitude of a quantitative property or relation to a unit of the same type of quantitative property or relation.
Variance	The variance of a random variable is a measure of its statistical dispersion, indicating how far from the expected value its values typically are. The variance of a real-valued random variable is its second central moment, and it also happens to be its second cumulant.
Multicolline-rity	Multicollinearity refers to linear inter-correlation among variables. Simply put, if nominally "different" measures actually quantify the same phenomenon to a significant degree -- i.e., wherein the variables are accorded different names and perhaps employ different numeric measurement scales but correlate highly with each other -- they are redundant.
Least Squares	Least squares is a mathematical optimization technique which, when given a series of measured data, attempts to find a function which closely approximates the data (a "best fit"). It attempts to minimize the sum of the squares of the ordinate differences (called residuals) between points generated by the function and corresponding points in the data.
Probability	Probability is the ratio of the number of favorable outcomes to the number of possible outcomes.
Linear regression	In statistics, linear regression is a method of estimating the conditional expected value of one variable y given the values of some other variable or variables x. The variable of interest, y, is conventionally called the "response variable".
Odds ratio	The odds ratio is a measure of effect size particularly important in Bayesian statistics and logistic regression. It is defined as the ratio of the odds of an event occurring in one group to the odds of it occurring in another group, or to a data-based estimate of that ratio.
Odds	In probability theory and statistics the odds in favor of an event or a proposition are the quantity $p / (1 - p)$, where p is the probability of the event or proposition. The logarithm of the odds is the logit of the probability.
Statistic	A statistic is a characteristic of a sample drawn from a population.
Distribution	A distribution is a list of the values that a variable takes in a sample. It is usually a list, ordered by quantity.
Critical value	A critical value is the value corresponding to a given significance level. This cutoff value determines the boundary between those samples resulting in a test statistic that leads to rejecting the null hypothesis and those lead to a decision not to reject the null hypothesis.
Proportional	Two quantities are called proportional if they vary in such a way that one of the quantities is a constant multiple of the other, or equivalently if they have a constant ratio.
Regression coefficient	The regression coefficient is the slope of the straight line that most closely relates two correlated variables.
Nominal	In nominal measurement numerals are assigned to objects as labels or names. If two entities have the same number associated with them, they belong to the same category, and that is the only significance that they have. The only comparisons that can be made between variable values are equality and inequality.
A priori	In statistics, a priori knowledge refers to a knowledge of the actual population, rather than that estimated by observation.

Univariate	In statistics, in univariate data, each data point has only one scalar component. Or, when the statistical technique to be used contains only one dependent variable.
Predictor variable	The predictor variable is manipulated by the experimenter. By attempting to isolate all other factors, one can determine the influence of the independent variable on the dependent variable.
Multivariate analysis of variance	Multivariate analysis of variance is an extension of analysis of variance (ANOVA) methods to cover cases where there is more than one dependent variable and where the dependent variables cannot simply be combined.
Multivariate analysis	Multivariate analysis in statistics describes a collection of procedures which involve observation and analysis of more than one statistical variable at a time.
Analysis of variance	Analysis of variance (ANOVA) is a collection of statistical models and their associated procedures which compare means by splitting the overall observed variance into different parts.
Deviation	A deviation is the difference between an observed value and the expected value of a variable or function.
Standard deviation	The standard deviation is the most commonly used measure of statistical dispersion. Simply put, it measures how spread out the values in a data set are. The standard deviation is defined as the square root of the variance.
Scatterplot	A scatterplot is used in statistics to visually display and compare two or more sets of related quantitative, or numerical, data by displaying only finitely many points, each having a coordinate on a horizontal and a vertical axis.
Combinations	Combinations are un-ordered collections of unique elements. The order of the elements is not important
Sample size	Sensitivity can be increased by using statistical controls, by increasing the reliability of measures (as in psychometric reliability), and by increasing the size of the sample. Increasing sample size is the most commonly used method for increasing statistical power.
Dispersion	In descriptive statistics, statistical dispersion is quantifiable variation of measurements of differing members of a population within the scale on which they are measured.
Subset	A is a subset of a set B, if A is "contained" inside B. The relationship of one set being a subset of another is called inclusion. Every set is a subset of itself.
Exhaustive	In set theory, exhaustive is an attribute of rule(s) for defining the inclusion/exclusion of elements of a collection of sets such that no non-exclusive elements remain which cannot definitely be included in one and only one set.
Mutually exclusive	In probability theory, events E_1, E_2, ..., E_n are said to be mutually exclusive if the occurrence of any one them automatically implies the non-occurrence of the remaining $n - 1$ events. In other words, two mutually exclusive events cannot both occur.
Sampling	Sampling is that part of statistical practice concerned with the selection of individual observations intended to yield some knowledge about a population of concern, especially for the purposes of statistical inference.
Stratified sampling	When sub-populations vary considerably, it is advantageous to sample each subpopulation independently. Stratified sampling is the process of grouping members of the population into relatively homogeneous subgroups before sampling.
Outlier	In statistics, an outlier is a single observation "far away" from the rest of the data. In most samplings of data, some data points will be further away from their expected values than what is deemed reasonable. This can be due to systematic error or faults in the theory that

Go to **Cram101.com** for the Practice Tests for this Chapter.

	generated the expected values.
Stepwise Regression	Stepwise regression is an incremental approach to multiple regression where the order of entry of predictor variables is determined systematically.
Significance level	The significance level of a test is the maximum probability of accidentally rejecting a true null hypothesis (a decision known as a Type I error). The significance of a result is also called its p-value; the smaller the p-value, the more significant the result is said to be.
Construct	A construct is a mathematical or conceptual model.
Regression equation	The regression equation represents the relation between selected values of one variable (x) and observed values of the other (y); it permits the prediction of the most probable values of y.
Population	A population is a set of entities concerning which statistical inferences are to be drawn, often based on a random sample taken from the population.
Weighted average	In statistics, given a set of data, X = { x1, x2, ..., xn} and corresponding non-negative weights, W = { w1, w2, ..., wn} the weighted average, is calculated as: Mean = $\sum$ w.x. / $\sum$ w.
Bias	A bias is a prejudice in a general or specific sense, usually in the sense for having a predilection to one particular point of view or ideological perspective. However, one is generally only said to be biased if one's powers of judgement are influenced by the biases one holds, to the extent that one's views could not be taken as being neutral or objective, but instead as subjective.
Upper limit	The upper limit in a distribution is the greatest score or value in the distribution.
Confidence Level	The confidence level is the range (with a specified value of uncertainty, usually expressed in percent) within which the true value of a measured quantity exists.
Residual	Error is a misnomer; an error is the amount by which an observation differs from its expected value; the latter being based on the whole population from which the statistical unit was chosen randomly. A residual, on the other hand, is an observable estimate of the unobservable error.
Origin	The point of intersection, where the axes meet, is called the origin normally labeled O. With the origin labeled O, we can name the x axis Ox and the y axis Oy. The x and y axes define a plane that can be referred to as the xy plane.
Range	In descriptive statistics, the range is the length of the smallest interval which contains all the data. It is calculated by subtracting the smallest observations from the greatest and provides an indication of statistical dispersion.
Slope	The slope is commonly used to describe the measurement of the steepness, incline or grade of a straight line. A higher slope value indicates a steeper incline.
Goodness of fit	Goodness of fit means how well a statistical model fits a set of observations. Measures of goodness of fit typically summarize the discrepancy between observed values and the values expected under the model in question. Such measures can be used in statistical hypothesis testing, e.g. to test for normality of residuals, to test whether two samples are drawn from identical distributions.
Total sum of squares	The total sum of squares is the sum of the squares of the difference of the independent variable and its grand mean. Furthermore, total sum of squares = explained sum of squares + residual sum of squares.
Baseline	Baseline is a starting point or condition against which future changes are measured.

Sum of squares	Sum of squares is a concept that permeates much of inferential statistics and descriptive statistics. More properly, it is "the sum of the squared deviations". Mathematically, it is an unscaled, or unadjusted measure of dispersion. When scaled for the number of degrees of freedom, it becomes the variance, the sum of squares per degree of freedom.
Coefficient of determination	The coefficient of determination R^2 is the proportion of a sample variance of a response variable that is "explained" by the predictor variables when a linear regression is done.
Hypothesis	A hypothesis is a proposed explanation for a phenomenon. A scientific hypothesis must be testable and based on previous observations or extensions of scientific theories.
Hypothesis testing	Hypothesis testing is an algorithm to state the alternative which minimizes certain risks.
Dichotomous variable	A variable with only two possible values is a dichotomous variable.
Descriptive statistics	Descriptive statistics is a branch of statistics that denotes any of the many techniques used to summarize a set of data. In a sense, we are using the data on members of a set to describe the set.
Degrees of freedom	In fitting statistical models to data, the vectors of residuals are often constrained to lie in a space of smaller dimension than the number of components in the vector. That smaller dimension is the number of degrees of freedom for error.
Sets	Sets are collections of objects considered as a whole. The objects of sets are called elements or members. The elements of a set can be anything: numbers, people, letters of the alphabet, other sets, and so on. Sets are conventionally denoted with capital letters, A, B, C, etc. Two sets A and B are said to be equal, written A = B, if they have the same members.
Pooled variance	Under the assumption of equal population variances, the pooled variance represents the best estimate of this equal but unknown population variance. It is a weighted average of the variance within each group.
Negative relationship	Negative relationship occurs when a change in one variable results in or is correlated with a change in another variable in the opposite direction. If the first increases, the second decreases, and so on.
Test Statistic	A test statistic is a summary value (often a summary statistic) of a data set that is compared with a statistical distribution to determine whether the data set differs from that expected under a null hypothesis.
Partial correlation	A partial correlation is a correlation between two variables when the effects of one or more related variables are removed.
Conditional probability	Conditional probability is the probability of some event A, given that some other event, B, has already occurred. Conditional probability is written $P(A \mid B)$, and is read "the probability of A, given B".
Standard error	The standard error of a measurement, value or quantity is the standard deviation of the process by which it was generated, after adjusting for sample size.
Statistical power	Statistical power depends on the significance criterion, the size of the difference or the strength of the similarity (that is, the effect size) in the population, and the sensitivity of the data.
Multivariate statistical analysis	Multivariate statistical analysis in statistics describes a collection of procedures which involve observation and analysis of more than one statistical variable at a time.
Statistical	Statistical analysis refers to the branch of mathematics that deals with the collection,

analysis	analysis, interpretation and presentation of masses of numerical data.
Criterion variable	The variable in a study that is expected to change as a result of alteration of the independent variable is the criterion variable.
Parameter	A parameter is a characteristic of a population.
Random sample	A sample is a subset chosen from a population for investigation. A random sample is one chosen by a method involving an unpredictable component, in the sense that the selection of any element of the population is independent of the selection of any other element.

Go to **Cram101.com** for the Practice Tests for this Chapter.

Multivariate analysis of variance	Multivariate analysis of variance is an extension of analysis of variance (ANOVA) methods to cover cases where there is more than one dependent variable and where the dependent variables cannot simply be combined.
Analysis of covariance	Analysis of covariance is an old-fashioned name for a linear regression model with one continuous explanatory variable and one or more factors. The name exists for historical reasons, but there is no particular reason to distinguish the method from the general purpose linear model.
Multivariate analysis	Multivariate analysis in statistics describes a collection of procedures which involve observation and analysis of more than one statistical variable at a time.
Independent variable	An independent variable is presumed to cause or determine a dependent variable. It can be changed as required and its values do not represent a problem requiring explanation in an analysis, but are taken simply as given.
Analysis of variance	Analysis of variance (ANOVA) is a collection of statistical models and their associated procedures which compare means by splitting the overall observed variance into different parts.
Multivariate	In statistics, in multivariate data, each data point has more than one scalar component, and often one is concerned with correlations between the components.
Interaction	Interaction is a kind of action which occurs as two or more objects have an effect upon one another. The idea of a two-way effect is essential in the concept of interaction instead of a one-way causal effect.
Covariance	Intuitively, covariance is the measure of how much two variables vary together. That is to say, the covariance becomes more positive for each pair of values which differ from their mean in the same direction, and becomes more negative with each pair of values which differ from their mean in opposite directions.
Construct	A construct is a mathematical or conceptual model.
Statistic	A statistic is a characteristic of a sample drawn from a population.
Variable	A variable is a symbol denoting a quantity or symbolic representation. In mathematics, a variable often represents an unknown quantity; in computer science, it represents a place where a quantity can be stored.
Variance	The variance of a random variable is a measure of its statistical dispersion, indicating how far from the expected value its values typically are. The variance of a real-valued random variable is its second central moment, and it also happens to be its second cumulant.
Test Statistic	A test statistic is a summary value (often a summary statistic) of a data set that is compared with a statistical distribution to determine whether the data set differs from that expected under a null hypothesis.
Dependent variable	A variable that changes as a function of a change to another variable is called a dependent variable.
Sampling Variability	Sampling variability is the variability of the estimate of a population characteristic due to sampling error.
Sampling	Sampling is that part of statistical practice concerned with the selection of individual observations intended to yield some knowledge about a population of concern, especially for the purposes of statistical inference.
Random sampling	In random sampling every combination of items from the frame, or stratum, has a known probability of occurring, but these probabilities are not necessarily equal. With any form of sampling there is a risk that the sample may not adequately represent the population but with

Go to **Cram101.com** for the Practice Tests for this Chapter.

random sampling there is a large body of statistical theory which quantifies the risk and thus enables an appropriate sample size to be chosen.

Significance level	The significance level of a test is the maximum probability of accidentally rejecting a true null hypothesis (a decision known as a Type I error). The significance of a result is also called its p-value; the smaller the p-value, the more significant the result is said to be.
Type I error	A false positive, also called a Type I error, exists when a test incorrectly reports that it has found a result where none really exists.
Population	A population is a set of entities concerning which statistical inferences are to be drawn, often based on a random sample taken from the population.
Sample	A sample is that part of a population which is actually observed. In normal scientific practice, we demand that it is selected in such a way as to avoid presenting a biased view of the population.
Mean	For a real-valued random variable X, the mean is the expectation of X. If the expectation does not exist, then the random variable has no mean. For a data set, the mean is just the sum of all the observations divided by the number of observations.
Blocking	In the statistical theory of the design of experiments, blocking is the arranging of experimental units in groups (blocks) which are similar to one another.
Alpha level	Among all the sets of possible values, we must choose one that we think represents the most extreme evidence against the hypothesis. That is called the critical region of the test statistic. The probability of the test statistic falling in the critical region when the hypothesis is correct is called the alpha level for the test.
Bonferroni	The Bonferroni Correction states that if an experimenter is testing n independent hypotheses on a set of data, then the statistical significance level that should be used is n times smaller than usual. e.g. when testing two hypotheses, instead of a p value of 0.05, one would use a stricter p value of 0.025. The Bonferroni Correction is a safeguard against multiple tests of statistical significance on the same data.
Critical value	A critical value is the value corresponding to a given significance level. This cutoff value determines the boundary between those samples resulting in a test statistic that leads to rejecting the null hypothesis and those lead to a decision not to reject the null hypothesis.
Discriminant	A discriminant is an expression that discriminates qualities of algebraic structures. The concept applies to polynomials, conic sections, quadratic forms, and algebraic number fields.
Disordinal interaction	In the disordinal interaction, regression lines cross, and in some cases, one treatment is better than a second, but in some other cases the first is worse than the second.
Main effect	The main effect is the direct effect that a each independent variable has on the dependent variable without regard to the possibility of interactions.
Power	The power of a statistical test is the probability that the test will reject a false null hypothesis, or in other words that it will not make a Type II error. The higher the power, the greater the chance of obtaining a statistically significant result when the null hypothesis is false.
Effect size	An effect size describes how large the relationship is between two variables. This information is important in scientific research. Often it is useful to know not only whether an experiment had an effect, but also the size of any effects.
Statistical power	Statistical power depends on the significance criterion, the size of the difference or the strength of the similarity (that is, the effect size) in the population, and the sensitivity of the data.

Deviation	A deviation is the difference between an observed value and the expected value of a variable or function.
Standard deviation	The standard deviation is the most commonly used measure of statistical dispersion. Simply put, it measures how spread out the values in a data set are. The standard deviation is defined as the square root of the variance.
Predictor variable	The predictor variable is manipulated by the experimenter. By attempting to isolate all other factors, one can determine the influence of the independent variable on the dependent variable.
Combinations	Combinations are un-ordered collections of unique elements. The order of the elements is not important
Factorial	The factorial of a natural number n is the product of all positive integers less than and equal to n. This is written as n! and pronounced "n factorial". The notation n! was introduced by Christian Kramp in 1808.
Factorial design	A factorial design is a statistical study in which each observation is categorized according to more than one factor. Such an experiment allows studying the effect of each factor on the response variable, while requiring fewer observations than by conducting separate experiments for each factor independently. It also allows studying the effect of the interaction between factors on the response variable.
Hypothesis	A hypothesis is a proposed explanation for a phenomenon. A scientific hypothesis must be testable and based on previous observations or extensions of scientific theories.
Null hypothesis	A null hypothesis, H_0, is a hypothesis set up to be nullified or refuted in order to support an alternative hypothesis. When used, the null hypothesis is presumed true until statistical evidence in the form of a hypothesis test indicates otherwise.
Statistical significance	A result is has statistical significance if it is unlikely to have occurred by chance, given that a presumed null hypothesis is true.
Normal distribution	The normal distribution is an extremely important probability distribution in many fields. It is a family of distributions of the same general form, differing in their location and scale parameters: the mean and standard deviation. The standard normal distribution is the normal distribution with a mean of zero and a standard deviation of one
Distribution	A distribution is a list of the values that a variable takes in a sample. It is usually a list, ordered by quantity.
Univariate	In statistics, in univariate data, each data point has only one scalar component. Or, when the statistical technique to be used contains only one dependent variable.
Generalization	Concept A is a (strict) generalization of concept B if and only if: every instance of concept B is also an instance of concept A; and there are instances of concept A which are not instances of concept B.
Ordinal interaction	An ordinal interaction has the property that throughout the range under consideration, one treatment is always superior to the other. In such an interaction, the regression lines do not cross.
Ordinal	In ordinal measurement the numbers assigned to objects represent the rank order (1st, 2nd, 3rd etc) of the entities measured. Comparisons of greater and less can be made, in addition to equality and inequality. However operations such as conventional addition and subtraction are still without meaning.
Association	In statistics, an association comes from two variables that are related. Many people confuse association with cauzation. Association does not imply cauzation.

Orthogonal	In mathematics, orthogonal is synonymous with perpendicular when used as a simple adjective that is not part of any longer phrase with a standard definition. It means at right angles.
A priori	In statistics, a priori knowledge refers to a knowledge of the actual population, rather than that estimated by observation.
Confidence Level	The confidence level is the range (with a specified value of uncertainty, usually expressed in percent) within which the true value of a measured quantity exists.
Probability	Probability is the ratio of the number of favorable outcomes to the number of possible outcomes.
Sample size	Sensitivity can be increased by using statistical controls, by increasing the reliability of measures (as in psychometric reliability), and by increasing the size of the sample. Increasing sample size is the most commonly used method for increasing statistical power.
Repeated measures	An experiment in which the same subjects are assigned to each group is called repeated measures. Such an experimental design is intended to show time effects, exposure effects, or to reduce variability.
Replication	Replication is repeating the creation of a phenomenon, so that the variability associated with the phenomenon can be estimated.
Dispersion	In descriptive statistics, statistical dispersion is quantifiable variation of measurements of differing members of a population within the scale on which they are measured.
Standard error	The standard error of a measurement, value or quantity is the standard deviation of the process by which it was generated, after adjusting for sample size.
Discriminant analysis	Linear discriminant analysis and the related Fisher's linear discriminant are used in machine learning to find the linear combination of features which best separate two or more classes of object or event.
Stepwise Regression	Stepwise regression is an incremental approach to multiple regression where the order of entry of predictor variables is determined systematically.
Alternative hypothesis	The alternate hypothesis, or alternative hypothesis, together with the null hypothesis are the two rival hypothesis whose likelihoods are compared by a statistical hypothesis test. Usually the alternate hypothesis is the possibility that an observed effect is genuine and the null hypothesis is the rival possibility that it has resulted from random chance.
Population mean	The expected value of a random variable is called the population mean.
Type II error	In statistics, a false negative, also called a Type II error or miss, exists when a test incorrectly reports that a result was not detected, when it was really present.
Multiple regression	A multiple regression is a linear regression with more than one covariate (predictor variable). It can be viewed as a simple case of canonical correlation. An equation used to predict a dependent variable, y from two independents, u and v is: $y = \beta_0 + \beta_1 u + \beta_2 v + \beta_3 u2 + \beta_4 uv + \beta_5 v2$
Sample Mean	The arithmetic mean of a set of numbers is the sum of all the members of the set divided by the number of items in the set. If the set is a statistical population, then we speak of the population mean; if of a sampling of a population, it is a sample mean.
Sampling error	When analyzing collected data, the samples observed differ in such things as means and standard deviations from the population from which the sample is taken. This is sampling error and is controlled by ensuring that, as much as possible, the samples taken have no systematic characteristics and are a true random sample from all possible samples.
Degrees of	In fitting statistical models to data, the vectors of residuals are often constrained to lie

freedom	in a space of smaller dimension than the number of components in the vector. That smaller dimension is the number of degrees of freedom for error.
Mean square within	The mean square within in an ANOVA is the sum of the squared deviations within the subject scores regardless of treatment group, relative to the number of degrees of freedom for the number of subjects.
Mean square between	The mean square between in an ANOVA is the sum of the squared deviations of a treatment group with respect to the number of degrees of freedom for that treatement group.
Population variance	If $\mu = E(X)$ is the expected value (mean) of the random variable X, then the population variance is $var(X) = E((X - \mu)^2)$.
Expected value	The expected value (or mathematical expectation) of a random variable is the sum of the probability of each possible outcome of the experiment multiplied by its payoff ("value").
Statistical Inference	Statistical inference is inference about a population from a random sample drawn from it or, more generally, about a random process from its observed behavior during a finite period of time. It includes: point estimation, interval estimation, hypothesis testing (or statistical significance testing) prediction
Hypothesis testing	Hypothesis testing is an algorithm to state the alternative which minimizes certain risks.
Range	In descriptive statistics, the range is the length of the smallest interval which contains all the data. It is calculated by subtracting the smallest observations from the greatest and provides an indication of statistical dispersion.
Correlation	Correlation indicates the strength and direction of a linear relationship between two random variables. In general statistical usage, correlation refers to the departure of two variables from independence.
Multicollinerity	Multicollinearity refers to linear inter-correlation among variables. Simply put, if nominally "different" measures actually quantify the same phenomenon to a significant degree -- i.e., wherein the variables are accorded different names and perhaps employ different numeric measurement scales but correlate highly with each other -- they are redundant.
Generalizability	Generalizability describes whether the results of individual studies and investigation samples can be applied to other studies and investigation samples.
Outlier	In statistics, an outlier is a single observation "far away" from the rest of the data. In most samplings of data, some data points will be further away from their expected values than what is deemed reasonable. This can be due to systematic error or faults in the theory that generated the expected values.
Subset	A is a subset of a set B, if A is "contained" inside B. The relationship of one set being a subset of another is called inclusion. Every set is a subset of itself.
Regression analysis	Regression analysis is any statistical method where the mean of one or more random variables is predicted conditioned on other (measured) random variables.
Linear regression	In statistics, linear regression is a method of estimating the conditional expected value of one variable y given the values of some other variable or variables x. The variable of interest, y, is conventionally called the "response variable".
Residual	Error is a misnomer; an error is the amount by which an observation differs from its expected value; the latter being based on the whole population from which the statistical unit was chosen randomly. A residual, on the other hand, is an observable estimate of the unobservable error.
Bias	A bias is a prejudice in a general or specific sense, usually in the sense for having a

predilection to one particular point of view or ideological perspective. However, one is generally only said to be biased if one's powers of judgement are influenced by the biases one holds, to the extent that one's views could not be taken as being neutral or objective, but instead as subjective.

Random assignment

In experimental design, the random assignment, or random placement of participants in experimental versus control groups in order to ensure that any differences between/among the groups are not systematic at the outset of the experiment. Random assignment does not ensure that the groups are "matched" or equivalent, only that any differences are due to chance.

Systematic bias

An example of systematic bias would be a thermometer that always read three degrees colder than the actual temperature because of incorrect initial calibration or labelling, whereas one that gave random values within five degrees either side of the actual temperature would have random error.

Elements

A set is a collection of objects considered as a whole. The objects of a set are called elements or members. The elements of a set can be anything: numbers, people, letters of the alphabet, other sets, and so on.

Skewness

Skewness is a measure of the asymmetry of the distribution of a real-valued random variable. Skewness, the third standardized moment, is written as $\gamma1$ and defined as $\gamma1 = \mu^3 / \sigma^3$ where μ^3 is the third moment about the mean and σ is the standard deviation.

Linear relationship

A situation in which the best-fitting regression line is a straight line is called a linear relationship.

Reliability

Reliability has been defined in different ways by different authors. Perhaps the best way to look at reliability is the extent to which the measurements resulting from a test are the result of characteristics of those being measured.

Regression equation

The regression equation represents the relation between selected values of one variable (x) and observed values of the other (y); it permits the prediction of the most probable values of y.

Tukey HSD

To test all pairwise comparisons among means using the Tukey HSD, compute t for each pair of means using the formula: $t_c = (M_i - M_j) / \sqrt{(MSE/n_h)}$ where $M_i - M_j$ is the difference between the ith and jth means, MSE is the Mean Square Error, and n_h is the harmonic mean of the sample sizes of groups i and j.

Homogeneity of variance

Homogeneity of variance is the situation in which two or more populations have or are assumed to have equal variances.

Multivariate statistical analysis

Multivariate statistical analysis in statistics describes a collection of procedures which involve observation and analysis of more than one statistical variable at a time.

Statistical analysis

Statistical analysis refers to the branch of mathematics that deals with the collection, analysis, interpretation and presentation of masses of numerical data.

Frequency

Frequency is the measurement of the number of times that a repeated event occurs per unit of time.

Variable	A variable is a symbol denoting a quantity or symbolic representation. In mathematics, a variable often represents an unknown quantity; in computer science, it represents a place where a quantity can be stored.
Predictor variable	The predictor variable is manipulated by the experimenter. By attempting to isolate all other factors, one can determine the influence of the independent variable on the dependent variable.
Combinations	Combinations are un-ordered collections of unique elements. The order of the elements is not important
Interaction	Interaction is a kind of action which occurs as two or more objects have an effect upon one another. The idea of a two-way effect is essential in the concept of interaction instead of a one-way causal effect.
Main effect	The main effect is the direct effect that a each independent variable has on the dependent variable without regard to the possibility of interactions.
Independent variable	An independent variable is presumed to cause or determine a dependent variable. It can be changed as required and its values do not represent a problem requiring explanation in an analysis, but are taken simply as given.
Measurement	Measurement generally refers to the process of estimating or determining the ratio of a magnitude of a quantitative property or relation to a unit of the same type of quantitative property or relation.
Dependent variable	A variable that changes as a function of a change to another variable is called a dependent variable.
Multivariate	In statistics, in multivariate data, each data point has more than one scalar component, and often one is concerned with correlations between the components.
Discriminant analysis	Linear discriminant analysis and the related Fisher's linear discriminant are used in machine learning to find the linear combination of features which best separate two or more classes of object or event.
Discriminant	A discriminant is an expression that discriminates qualities of algebraic structures. The concept applies to polynomials, conic sections, quadratic forms, and algebraic number fields.
Regression analysis	Regression analysis is any statistical method where the mean of one or more random variables is predicted conditioned on other (measured) random variables.
Orthogonal	In mathematics, orthogonal is synonymous with perpendicular when used as a simple adjective that is not part of any longer phrase with a standard definition. It means at right angles.
Range	In descriptive statistics, the range is the length of the smallest interval which contains all the data. It is calculated by subtracting the smallest observations from the greatest and provides an indication of statistical dispersion.
Factorial	The factorial of a natural number n is the product of all positive integers less than and equal to n. This is written as n! and pronounced "n factorial". The notation n! was introduced by Christian Kramp in 1808.
Factorial design	A factorial design is a statistical study in which each observation is categorized according to more than one factor. Such an experiment allows studying the effect of each factor on the response variable, while requiring fewer observations than by conducting separate experiments for each factor independently. It also allows studying the effect of the interaction between factors on the response variable.
Subset	A is a subset of a set B, if A is "contained" inside B. The relationship of one set being a subset of another is called inclusion. Every set is a subset of itself.

Correlation	Correlation indicates the strength and direction of a linear relationship between two random variables. In general statistical usage, correlation refers to the departure of two variables from independence.
Deviation	A deviation is the difference between an observed value and the expected value of a variable or function.
Reliability	Reliability has been defined in different ways by different authors. Perhaps the best way to look at reliability is the extent to which the measurements resulting from a test are the result of characteristics of those being measured.
Sample	A sample is that part of a population which is actually observed. In normal scientific practice, we demand that it is selected in such a way as to avoid presenting a biased view of the population.
Construct	A construct is a mathematical or conceptual model.
Mean	For a real-valued random variable X, the mean is the expectation of X. If the expectation does not exist, then the random variable has no mean. For a data set, the mean is just the sum of all the observations divided by the number of observations.
Squared deviations	We use the deviations about the mean in many analyses to determine the amount of spread or distance among the scores to determine whether their is a significant enough difference to warrant saying that the values all describe the same object or whether they most likely describe different objects. Since we are interested in distance and not direction, we need the absolute values of the deviations. We use squared deviations which converts all numbers to positive values. Taking the square root of the result gives an absolute value
Sets	Sets are collections of objects considered as a whole. The objects of sets are called elements or members. The elements of a set can be anything: numbers, people, letters of the alphabet, other sets, and so on. Sets are conventionally denoted with capital letters, A, B, C, etc. Two sets A and B are said to be equal, written A = B, if they have the same members.
Statistical analysis	Statistical analysis refers to the branch of mathematics that deals with the collection, analysis, interpretation and presentation of masses of numerical data.
Multicolline-rity	Multicollinearity refers to linear inter-correlation among variables. Simply put, if nominally "different" measures actually quantify the same phenomenon to a significant degree -- i.e., wherein the variables are accorded different names and perhaps employ different numeric measurement scales but correlate highly with each other -- they are redundant.
Parameter	A parameter is a characteristic of a population.
A priori	In statistics, a priori knowledge refers to a knowledge of the actual population, rather than that estimated by observation.
Power	The power of a statistical test is the probability that the test will reject a false null hypothesis, or in other words that it will not make a Type II error. The higher the power, the greater the chance of obtaining a statistically significant result when the null hypothesis is false.
Variance	The variance of a random variable is a measure of its statistical dispersion, indicating how far from the expected value its values typically are. The variance of a real-valued random variable is its second central moment, and it also happens to be its second cumulant.
Linear relationship	A situation in which the best-fitting regression line is a straight line is called a linear relationship.
Regression coefficient	The regression coefficient is the slope of the straight line that most closely relates two correlated variables.

Confounded	Confounded is the situation in which the effect of a controlled variable is inextricably mixed with that of another, uncontrolled variable.
Goodness of fit	Goodness of fit means how well a statistical model fits a set of observations. Measures of goodness of fit typically summarize the discrepancy between observed values and the values expected under the model in question. Such measures can be used in statistical hypothesis testing, e.g. to test for normality of residuals, to test whether two samples are drawn from identical distributions.
Analysis of variance	Analysis of variance (ANOVA) is a collection of statistical models and their associated procedures which compare means by splitting the overall observed variance into different parts.
Ordinal	In ordinal measurement the numbers assigned to objects represent the rank order (1st, 2nd, 3rd etc) of the entities measured. Comparisons of greater and less can be made, in addition to equality and inequality. However operations such as conventional addition and subtraction are still without meaning.
Multiple regression	A multiple regression is a linear regression with more than one covariate (predictor variable). It can be viewed as a simple case of canonical correlation. An equation used to predict a dependent variable, y from two independents, u and v is: $y = \beta_0 + \beta_1 u + \beta_2 v + \beta_3 u2 + \beta_4 uv + \beta_5 v2$
Statistical significance	A result is has statistical significance if it is unlikely to have occurred by chance, given that a presumed null hypothesis is true.
Population	A population is a set of entities concerning which statistical inferences are to be drawn, often based on a random sample taken from the population.
Sampling	Sampling is that part of statistical practice concerned with the selection of individual observations intended to yield some knowledge about a population of concern, especially for the purposes of statistical inference.
Sampling error	When analyzing collected data, the samples observed differ in such things as means and standard deviations from the population from which the sample is taken. This is sampling error and is controlled by ensuring that, as much as possible, the samples taken have no systematic characteristics and are a true random sample from all possible samples.
Complement	Generally a complement of X is something that together with X makes a complete whole; that supplies what X lacks.
Probability	Probability is the ratio of the number of favorable outcomes to the number of possible outcomes.
Distribution	A distribution is a list of the values that a variable takes in a sample. It is usually a list, ordered by quantity.
Origin	The point of intersection, where the axes meet, is called the origin normally labeled O. With the origin labeled O, we can name the x axis Ox and the y axis Oy. The x and y axes define a plane that can be referred to as the xy plane.
Elements	A set is a collection of objects considered as a whole. The objects of a set are called elements or members. The elements of a set can be anything: numbers, people, letters of the alphabet, other sets, and so on.
Mode	The mode is the value that has the largest number of observations, namely the most frequent value within a particular set of values.
Association	In statistics, an association comes from two variables that are related. Many people confuse association with cauzation. Association does not imply cauzation.

| **Statistic** | A statistic is a characteristic of a sample drawn from a population. |
| **Sample size** | Sensitivity can be increased by using statistical controls, by increasing the reliability of measures (as in psychometric reliability), and by increasing the size of the sample. Increasing sample size is the most commonly used method for increasing statistical power. |

Variable	A variable is a symbol denoting a quantity or symbolic representation. In mathematics, a variable often represents an unknown quantity; in computer science, it represents a place where a quantity can be stored.
Sets	Sets are collections of objects considered as a whole. The objects of sets are called elements or members. The elements of a set can be anything: numbers, people, letters of the alphabet, other sets, and so on. Sets are conventionally denoted with capital letters, A, B, C, etc. Two sets A and B are said to be equal, written A = B, if they have the same members.
Canonical correlation	Canonical correlation analysis seeks vectors a and b such that the random variables a'X and b'Y maximize the correlation p = cor(a'X, b'Y)..
Multiple regression	A multiple regression is a linear regression with more than one covariate (predictor variable). It can be viewed as a simple case of canonical correlation. An equation used to predict a dependent variable, y from two independents, u and v is: $y = \beta_0 + \beta_1 u + \beta_2 v + \beta_3 u2 + \beta_4 uv + \beta_5 v2$
Correlation	Correlation indicates the strength and direction of a linear relationship between two random variables. In general statistical usage, correlation refers to the departure of two variables from independence.
Multivariate	In statistics, in multivariate data, each data point has more than one scalar component, and often one is concerned with correlations between the components.
Multivariate analysis of variance	Multivariate analysis of variance is an extension of analysis of variance (ANOVA) methods to cover cases where there is more than one dependent variable and where the dependent variables cannot simply be combined.
Multivariate analysis	Multivariate analysis in statistics describes a collection of procedures which involve observation and analysis of more than one statistical variable at a time.
Independent variable	An independent variable is presumed to cause or determine a dependent variable. It can be changed as required and its values do not represent a problem requiring explanation in an analysis, but are taken simply as given.
Analysis of variance	Analysis of variance (ANOVA) is a collection of statistical models and their associated procedures which compare means by splitting the overall observed variance into different parts.
Variance	The variance of a random variable is a measure of its statistical dispersion, indicating how far from the expected value its values typically are. The variance of a real-valued random variable is its second central moment, and it also happens to be its second cumulant.
Dependent variable	A variable that changes as a function of a change to another variable is called a dependent variable.
Combinations	Combinations are un-ordered collections of unique elements. The order of the elements is not important
Orthogonal	In mathematics, orthogonal is synonymous with perpendicular when used as a simple adjective that is not part of any longer phrase with a standard definition. It means at right angles.
Regression equation	The regression equation represents the relation between selected values of one variable (x) and observed values of the other (y); it permits the prediction of the most probable values of y.
Discriminant analysis	Linear discriminant analysis and the related Fisher's linear discriminant are used in machine learning to find the linear combination of features which best separate two or more classes of object or event.
Discriminant	A discriminant is an expression that discriminates qualities of algebraic structures. The

concept applies to polynomials, conic sections, quadratic forms, and algebraic number fields.

Association	In statistics, an association comes from two variables that are related. Many people confuse association with cauzation. Association does not imply cauzation.
Sample size	Sensitivity can be increased by using statistical controls, by increasing the reliability of measures (as in psychometric reliability), and by increasing the size of the sample. Increasing sample size is the most commonly used method for increasing statistical power.
Sample	A sample is that part of a population which is actually observed. In normal scientific practice, we demand that it is selected in such a way as to avoid presenting a biased view of the population.
Range	In descriptive statistics, the range is the length of the smallest interval which contains all the data. It is calculated by subtracting the smallest observations from the greatest and provides an indication of statistical dispersion.
Statistical significance	A result is has statistical significance if it is unlikely to have occurred by chance, given that a presumed null hypothesis is true.
Distribution	A distribution is a list of the values that a variable takes in a sample. It is usually a list, ordered by quantity.
Univariate	In statistics, in univariate data, each data point has only one scalar component. Or, when the statistical technique to be used contains only one dependent variable.
Statistic	A statistic is a characteristic of a sample drawn from a population.
Bias	A bias is a prejudice in a general or specific sense, usually in the sense for having a predilection to one particular point of view or ideological perspective. However, one is generally only said to be biased if one's powers of judgement are influenced by the biases one holds, to the extent that one's views could not be taken as being neutral or objective, but instead as subjective.
Mean	For a real-valued random variable X, the mean is the expectation of X. If the expectation does not exist, then the random variable has no mean. For a data set, the mean is just the sum of all the observations divided by the number of observations.
Population	A population is a set of entities concerning which statistical inferences are to be drawn, often based on a random sample taken from the population.
Subset	A is a subset of a set B, if A is "contained" inside B. The relationship of one set being a subset of another is called inclusion. Every set is a subset of itself.
Sampling	Sampling is that part of statistical practice concerned with the selection of individual observations intended to yield some knowledge about a population of concern, especially for the purposes of statistical inference.
Sampling error	When analyzing collected data, the samples observed differ in such things as means and standard deviations from the population from which the sample is taken. This is sampling error and is controlled by ensuring that, as much as possible, the samples taken have no systematic characteristics and are a true random sample from all possible samples.
Test Statistic	A test statistic is a summary value (often a summary statistic) of a data set that is compared with a statistical distribution to determine whether the data set differs from that expected under a null hypothesis.
Multicolline-rity	Multicollinearity refers to linear inter-correlation among variables. Simply put, if nominally "different" measures actually quantify the same phenomenon to a significant degree -- i.e., wherein the variables are accorded different names and perhaps employ different numeric measurement scales but correlate highly with each other -- they are redundant.

Go to **Cram101.com** for the Practice Tests for this Chapter.

Regression analysis	Regression analysis is any statistical method where the mean of one or more random variables is predicted conditioned on other (measured) random variables.
Stepwise Regression	Stepwise regression is an incremental approach to multiple regression where the order of entry of predictor variables is determined systematically.
A priori	In statistics, a priori knowledge refers to a knowledge of the actual population, rather than that estimated by observation.
Construct	A construct is a mathematical or conceptual model.
Frequency	Frequency is the measurement of the number of times that a repeated event occurs per unit of time.
Predictor variable	The predictor variable is manipulated by the experimenter. By attempting to isolate all other factors, one can determine the influence of the independent variable on the dependent variable.
Criterion variable	The variable in a study that is expected to change as a result of alteration of the independent variable is the criterion variable.
Random sample	A sample is a subset chosen from a population for investigation. A random sample is one chosen by a method involving an unpredictable component, in the sense that the selection of any element of the population is independent of the selection of any other element.
Multidimensi-nal scaling	Multidimensional scaling is a set of related statistical techniques often used in data visualization for exploring similarities or dissimilarities in data. An algorithm starts with a matrix of item-item similarities, then assigns a location of each item in a low-dimensional space, suitable for graphing or 3D visualization.
Categorical variable	In a categorical variable, numerals are assigned to objects as labels or names. If two entities have the same number associated with them, they belong to the same category, and that is the only significance that they have. The only comparisons that can be made between variable values are equality and inequality.

Go to **Cram101.com** for the Practice Tests for this Chapter.

Construct	A construct is a mathematical or conceptual model.
Population	A population is a set of entities concerning which statistical inferences are to be drawn, often based on a random sample taken from the population.
Variable	A variable is a symbol denoting a quantity or symbolic representation. In mathematics, a variable often represents an unknown quantity; in computer science, it represents a place where a quantity can be stored.
Mean	For a real-valued random variable X, the mean is the expectation of X. If the expectation does not exist, then the random variable has no mean. For a data set, the mean is just the sum of all the observations divided by the number of observations.
Criterion validity	In psychometrics, criterion validity is a measure of how well one variable or set of variables predicts an outcome based on information from other variables. These variables are often represented as "intermediate" and "ultimate" criteria.
Outlier	In statistics, an outlier is a single observation "far away" from the rest of the data. In most samplings of data, some data points will be further away from their expected values than what is deemed reasonable. This can be due to systematic error or faults in the theory that generated the expected values.
Association	In statistics, an association comes from two variables that are related. Many people confuse association with cauzation. Association does not imply cauzation.
Deviation	A deviation is the difference between an observed value and the expected value of a variable or function.
Standard deviation	The standard deviation is the most commonly used measure of statistical dispersion. Simply put, it measures how spread out the values in a data set are. The standard deviation is defined as the square root of the variance.
Multicolline- rity	Multicollinearity refers to linear inter-correlation among variables. Simply put, if nominally "different" measures actually quantify the same phenomenon to a significant degree -- i.e., wherein the variables are accorded different names and perhaps employ different numeric measurement scales but correlate highly with each other -- they are redundant.
Bias	A bias is a prejudice in a general or specific sense, usually in the sense for having a predilection to one particular point of view or ideological perspective. However, one is generally only said to be biased if one's powers of judgement are influenced by the biases one holds, to the extent that one's views could not be taken as being neutral or objective, but instead as subjective.
Sum of squares between	The sum of squares between is the sum of the squared deviations around a group or treatment mean within an ANOVA design.
Sum of squares	Sum of squares is a concept that permeates much of inferential statistics and descriptive statistics. More properly, it is "the sum of the squared deviations". Mathematically, it is an unscaled, or unadjusted measure of dispersion. When scaled for the number of degrees of freedom, it becomes the variance, the sum of squares per degree of freedom.
Sample	A sample is that part of a population which is actually observed. In normal scientific practice, we demand that it is selected in such a way as to avoid presenting a biased view of the population.
Multivariate	In statistics, in multivariate data, each data point has more than one scalar component, and often one is concerned with correlations between the components.
Data reduction	All of statistical analysis provides data reduction, taking large amounts of information and reducing it into a more understandable form. Data reduction in this sense always results in a

Go to **Cram101.com** for the Practice Tests for this Chapter.

loss of information.

Statistical Inference	Statistical inference is inference about a population from a random sample drawn from it or, more generally, about a random process from its observed behavior during a finite period of time. It includes: point estimation, interval estimation, hypothesis testing (or statistical significance testing) prediction
Elements	A set is a collection of objects considered as a whole. The objects of a set are called elements or members. The elements of a set can be anything: numbers, people, letters of the alphabet, other sets, and so on.
Scatterplot	A scatterplot is used in statistics to visually display and compare two or more sets of related quantitative, or numerical, data by displaying only finitely many points, each having a coordinate on a horizontal and a vertical axis.
Range	In descriptive statistics, the range is the length of the smallest interval which contains all the data. It is calculated by subtracting the smallest observations from the greatest and provides an indication of statistical dispersion.
Discriminant analysis	Linear discriminant analysis and the related Fisher's linear discriminant are used in machine learning to find the linear combination of features which best separate two or more classes of object or event.
Discriminant	A discriminant is an expression that discriminates qualities of algebraic structures. The concept applies to polynomials, conic sections, quadratic forms, and algebraic number fields.
Correlation	Correlation indicates the strength and direction of a linear relationship between two random variables. In general statistical usage, correlation refers to the departure of two variables from independence.
Probability	Probability is the ratio of the number of favorable outcomes to the number of possible outcomes.
Sets	Sets are collections of objects considered as a whole. The objects of sets are called elements or members. The elements of a set can be anything: numbers, people, letters of the alphabet, other sets, and so on. Sets are conventionally denoted with capital letters, A, B, C, etc. Two sets A and B are said to be equal, written A = B, if they have the same members.
Measurement	Measurement generally refers to the process of estimating or determining the ratio of a magnitude of a quantitative property or relation to a unit of the same type of quantitative property or relation.
Nominal	In nominal measurement numerals are assigned to objects as labels or names. If two entities have the same number associated with them, they belong to the same category, and that is the only significance that they have. The only comparisons that can be made between variable values are equality and inequality.
Ordinal	In ordinal measurement the numbers assigned to objects represent the rank order (1st, 2nd, 3rd etc) of the entities measured. Comparisons of greater and less can be made, in addition to equality and inequality. However operations such as conventional addition and subtraction are still without meaning.
Dispersion	In descriptive statistics, statistical dispersion is quantifiable variation of measurements of differing members of a population within the scale on which they are measured.
Z score	A z score is a dimensionless quantity derived by subtracting the population mean from an individual (raw) score and then dividing the difference by the population standard deviation:
Standard score	A standard score (z) is a dimensionless quantity derived by subtracting the population mean from an individual (raw) score and then dividing the difference by the population standard

Go to **Cram101.com** for the Practice Tests for this Chapter.

deviation: $z = (X-\mu) / \sigma$.

Multidimensi-nal scaling	Multidimensional scaling is a set of related statistical techniques often used in data visualization for exploring similarities or dissimilarities in data. An algorithm starts with a matrix of item-item similarities, then assigns a location of each item in a low-dimensional space, suitable for graphing or 3D visualization.
Parameter	A parameter is a characteristic of a population.
Disjoint	In mathematics, two sets are said to be disjoint if they have no element in common. For example, {1, 2, 3} and {4, 5, 6} are disjoint sets.
Combinations	Combinations are un-ordered collections of unique elements. The order of the elements is not important
Sample size	Sensitivity can be increased by using statistical controls, by increasing the reliability of measures (as in psychometric reliability), and by increasing the size of the sample. Increasing sample size is the most commonly used method for increasing statistical power.
Random sample	A sample is a subset chosen from a population for investigation. A random sample is one chosen by a method involving an unpredictable component, in the sense that the selection of any element of the population is independent of the selection of any other element.
Statistical significance	A result is has statistical significance if it is unlikely to have occurred by chance, given that a presumed null hypothesis is true.
Common sense	There are two general meanings to the term common sense in philosophy. One is a sense that is common to the others, and the other meaning is a sense of things that is common to humanity.
A priori	In statistics, a priori knowledge refers to a knowledge of the actual population, rather than that estimated by observation.
Mode	The mode is the value that has the largest number of observations, namely the most frequent value within a particular set of values.
Predictive validity	In psychometrics, predictive validity is the extent to which a scale predicts scores on some criterion measure.
Independent variable	An independent variable is presumed to cause or determine a dependent variable. It can be changed as required and its values do not represent a problem requiring explanation in an analysis, but are taken simply as given.
Dependent variable	A variable that changes as a function of a change to another variable is called a dependent variable.
Generalizability	Generalizability describes whether the results of individual studies and investigation samples can be applied to other studies and investigation samples.
Replication	Replication is repeating the creation of a phenomenon, so that the variability associated with the phenomenon can be estimated.
Multivariate analysis	Multivariate analysis in statistics describes a collection of procedures which involve observation and analysis of more than one statistical variable at a time.
Kappa	Cohen's kappa coefficient is a statistical measure of inter-rater reliability. It is generally thought to be a more robust measure than simple percent agreement calculation since kappa takes into account the agreement occurring by chance. Cohen's kappa measures the agreement between two raters who each classify N items into C mutually exclusive categories.

Multidimensi-nal scaling	Multidimensional scaling is a set of related statistical techniques often used in data visualization for exploring similarities or dissimilarities in data. An algorithm starts with a matrix of item-item similarities, then assigns a location of each item in a low-dimensional space, suitable for graphing or 3D visualization.
Correlation	Correlation indicates the strength and direction of a linear relationship between two random variables. In general statistical usage, correlation refers to the departure of two variables from independence.
Variance	The variance of a random variable is a measure of its statistical dispersion, indicating how far from the expected value its values typically are. The variance of a real-valued random variable is its second central moment, and it also happens to be its second cumulant.
Multivariate	In statistics, in multivariate data, each data point has more than one scalar component, and often one is concerned with correlations between the components.
Covariance	Intuitively, covariance is the measure of how much two variables vary together. That is to say, the covariance becomes more positive for each pair of values which differ from their mean in the same direction, and becomes more negative with each pair of values which differ from their mean in opposite directions.
Categorical variable	In a categorical variable, numerals are assigned to objects as labels or names. If two entities have the same number associated with them, they belong to the same category, and that is the only significance that they have. The only comparisons that can be made between variable values are equality and inequality.
Variable	A variable is a symbol denoting a quantity or symbolic representation. In mathematics, a variable often represents an unknown quantity; in computer science, it represents a place where a quantity can be stored.
Measurement	Measurement generally refers to the process of estimating or determining the ratio of a magnitude of a quantitative property or relation to a unit of the same type of quantitative property or relation.
Independent variable	An independent variable is presumed to cause or determine a dependent variable. It can be changed as required and its values do not represent a problem requiring explanation in an analysis, but are taken simply as given.
Dependent variable	A variable that changes as a function of a change to another variable is called a dependent variable.
Range	In descriptive statistics, the range is the length of the smallest interval which contains all the data. It is calculated by subtracting the smallest observations from the greatest and provides an indication of statistical dispersion.
Mean	For a real-valued random variable X, the mean is the expectation of X. If the expectation does not exist, then the random variable has no mean. For a data set, the mean is just the sum of all the observations divided by the number of observations.
Combinations	Combinations are un-ordered collections of unique elements. The order of the elements is not important
Elements	A set is a collection of objects considered as a whole. The objects of a set are called elements or members. The elements of a set can be anything: numbers, people, letters of the alphabet, other sets, and so on.
Origin	The point of intersection, where the axes meet, is called the origin normally labeled O. With the origin labeled O, we can name the x axis Ox and the y axis Oy. The x and y axes define a plane that can be referred to as the xy plane.

Discriminant analysis	Linear discriminant analysis and the related Fisher's linear discriminant are used in machine learning to find the linear combination of features which best separate two or more classes of object or event.
Discriminant	A discriminant is an expression that discriminates qualities of algebraic structures. The concept applies to polynomials, conic sections, quadratic forms, and algebraic number fields.
Univariate	In statistics, in univariate data, each data point has only one scalar component. Or, when the statistical technique to be used contains only one dependent variable.
Sample	A sample is that part of a population which is actually observed. In normal scientific practice, we demand that it is selected in such a way as to avoid presenting a biased view of the population.
Distribution	A distribution is a list of the values that a variable takes in a sample. It is usually a list, ordered by quantity.
Goodness of fit	Goodness of fit means how well a statistical model fits a set of observations. Measures of goodness of fit typically summarize the discrepancy between observed values and the values expected under the model in question. Such measures can be used in statistical hypothesis testing, e.g. to test for normality of residuals, to test whether two samples are drawn from identical distributions.
Multiple regression	A multiple regression is a linear regression with more than one covariate (predictor variable). It can be viewed as a simple case of canonical correlation. An equation used to predict a dependent variable, y from two independents, u and v is: $y = \beta_0 + \beta_1 u + \beta_2 v + \beta_3 u2 + \beta_4 uv + \beta_5 v2$
Population	A population is a set of entities concerning which statistical inferences are to be drawn, often based on a random sample taken from the population.
Generalizability	Generalizability describes whether the results of individual studies and investigation samples can be applied to other studies and investigation samples.
Nominal	In nominal measurement numerals are assigned to objects as labels or names. If two entities have the same number associated with them, they belong to the same category, and that is the only significance that they have. The only comparisons that can be made between variable values are equality and inequality.
Association	In statistics, an association comes from two variables that are related. Many people confuse association with cauzation. Association does not imply cauzation.
Frequency	Frequency is the measurement of the number of times that a repeated event occurs per unit of time.
Probability	Probability is the ratio of the number of favorable outcomes to the number of possible outcomes.
Joint probability	Joint probability is the probability of two events in conjunction. That is, it is the probability of both events together. The joint probability of A and B is written P(AB).
Exploratory data analysis	Exploratory data analysis is that part of statistical practice concerned with reviewing, communicating and using data where there is a low level of knowledge about its cause system.
Sampling	Sampling is that part of statistical practice concerned with the selection of individual observations intended to yield some knowledge about a population of concern, especially for the purposes of statistical inference.
Sets	Sets are collections of objects considered as a whole. The objects of sets are called elements or members. The elements of a set can be anything: numbers, people, letters of the alphabet, other sets, and so on. Sets are conventionally denoted with capital letters, A, B,

C, etc. Two sets A and B are said to be equal, written A = B, if they have the same members.

Outlier

In statistics, an outlier is a single observation "far away" from the rest of the data. In most samplings of data, some data points will be further away from their expected values than what is deemed reasonable. This can be due to systematic error or faults in the theory that generated the expected values.

Statistical analysis

Statistical analysis refers to the branch of mathematics that deals with the collection, analysis, interpretation and presentation of masses of numerical data.

Reliability

Reliability has been defined in different ways by different authors. Perhaps the best way to look at reliability is the extent to which the measurements resulting from a test are the result of characteristics of those being measured.

Statistic

A statistic is a characteristic of a sample drawn from a population.

Data reduction

All of statistical analysis provides data reduction, taking large amounts of information and reducing it into a more understandable form. Data reduction in this sense always results in a loss of information.

Multivariate analysis

Multivariate analysis in statistics describes a collection of procedures which involve observation and analysis of more than one statistical variable at a time.

Statistical analysis	Statistical analysis refers to the branch of mathematics that deals with the collection, analysis, interpretation and presentation of masses of numerical data.
Measurement	Measurement generally refers to the process of estimating or determining the ratio of a magnitude of a quantitative property or relation to a unit of the same type of quantitative property or relation.
Variable	A variable is a symbol denoting a quantity or symbolic representation. In mathematics, a variable often represents an unknown quantity; in computer science, it represents a place where a quantity can be stored.
Multivariate analysis of variance	Multivariate analysis of variance is an extension of analysis of variance (ANOVA) methods to cover cases where there is more than one dependent variable and where the dependent variables cannot simply be combined.
Multivariate analysis	Multivariate analysis in statistics describes a collection of procedures which involve observation and analysis of more than one statistical variable at a time.
Independent variable	An independent variable is presumed to cause or determine a dependent variable. It can be changed as required and its values do not represent a problem requiring explanation in an analysis, but are taken simply as given.
Analysis of variance	Analysis of variance (ANOVA) is a collection of statistical models and their associated procedures which compare means by splitting the overall observed variance into different parts.
Multivariate	In statistics, in multivariate data, each data point has more than one scalar component, and often one is concerned with correlations between the components.
Variance	The variance of a random variable is a measure of its statistical dispersion, indicating how far from the expected value its values typically are. The variance of a real-valued random variable is its second central moment, and it also happens to be its second cumulant.
Dependent variable	A variable that changes as a function of a change to another variable is called a dependent variable.
Bootstrapping	Bootstrapping is a statistical method for estimating the sampling distribution of an estimator by sampling with replacement from the original sample, most often with the purpose of deriving robust estimates of standard errors and confidence intervals of a population parameter like a mean, median, proportion, odds ratio, correlation coefficient or regression coefficient.
Parameter	A parameter is a characteristic of a population.
Standard error	The standard error of a measurement, value or quantity is the standard deviation of the process by which it was generated, after adjusting for sample size.
Association	In statistics, an association comes from two variables that are related. Many people confuse association with cauzation. Association does not imply cauzation.
Correlation	Correlation indicates the strength and direction of a linear relationship between two random variables. In general statistical usage, correlation refers to the departure of two variables from independence.
Construct	A construct is a mathematical or conceptual model.
Abstract	An abstract is a brief summary of a research article, thesis, review, conference proceeding or any in-depth analysis of a particular subject or discipline, and is often used to help the reader quickly ascertain the paper's purpose.
Reliability	Reliability has been defined in different ways by different authors. Perhaps the best way to

look at reliability is the extent to which the measurements resulting from a test are the result of characteristics of those being measured.

Range	In descriptive statistics, the range is the length of the smallest interval which contains all the data. It is calculated by subtracting the smallest observations from the greatest and provides an indication of statistical dispersion.
Sampling distribution	A sampling distribution is the probability distribution, under repeated sampling of the population, of a given statistic (a numerical quantity calculated from the data values in a sample).
Distribution	A distribution is a list of the values that a variable takes in a sample. It is usually a list, ordered by quantity.
Sampling	Sampling is that part of statistical practice concerned with the selection of individual observations intended to yield some knowledge about a population of concern, especially for the purposes of statistical inference.
Degrees of freedom	In fitting statistical models to data, the vectors of residuals are often constrained to lie in a space of smaller dimension than the number of components in the vector. That smaller dimension is the number of degrees of freedom for error.
Covariance	Intuitively, covariance is the measure of how much two variables vary together. That is to say, the covariance becomes more positive for each pair of values which differ from their mean in the same direction, and becomes more negative with each pair of values which differ from their mean in opposite directions.
Confidence Interval	A confidence interval is an interval between two numbers, where there is a certain specified level of confidence that a population parameter lies.
Sampling error	When analyzing collected data, the samples observed differ in such things as means and standard deviations from the population from which the sample is taken. This is sampling error and is controlled by ensuring that, as much as possible, the samples taken have no systematic characteristics and are a true random sample from all possible samples.
Complement	Generally a complement of X is something that together with X makes a complete whole; that supplies what X lacks.
Sample	A sample is that part of a population which is actually observed. In normal scientific practice, we demand that it is selected in such a way as to avoid presenting a biased view of the population.
Confidence Level	The confidence level is the range (with a specified value of uncertainty, usually expressed in percent) within which the true value of a measured quantity exists.
Multiple regression	A multiple regression is a linear regression with more than one covariate (predictor variable). It can be viewed as a simple case of canonical correlation. An equation used to predict a dependent variable, y from two independents, u and v is: $y = \beta_0 + \beta_1 u + \beta_2 v + \beta_3 u2 + \beta_4 uv + \beta_5 v2$
Least Squares	Least squares is a mathematical optimization technique which, when given a series of measured data, attempts to find a function which closely approximates the data (a "best fit"). It attempts to minimize the sum of the squares of the ordinate differences (called residuals) between points generated by the function and corresponding points in the data.
Mean	For a real-valued random variable X, the mean is the expectation of X. If the expectation does not exist, then the random variable has no mean. For a data set, the mean is just the sum of all the observations divided by the number of observations.
Statistic	A statistic is a characteristic of a sample drawn from a population.

Go to **Cram101.com** for the Practice Tests for this Chapter.

Baseline	Baseline is a starting point or condition against which future changes are measured.
Generalizability	Generalizability describes whether the results of individual studies and investigation samples can be applied to other studies and investigation samples.
Predictor variable	The predictor variable is manipulated by the experimenter. By attempting to isolate all other factors, one can determine the influence of the independent variable on the dependent variable.
Ordinal	In ordinal measurement the numbers assigned to objects represent the rank order (1st, 2nd, 3rd etc) of the entities measured. Comparisons of greater and less can be made, in addition to equality and inequality. However operations such as conventional addition and subtraction are still without meaning.
Canonical correlation	Canonical correlation analysis seeks vectors a and b such that the random variables a'X and b'Y maximize the correlation p = cor(a'X, b'Y)..
Regression equation	The regression equation represents the relation between selected values of one variable (x) and observed values of the other (y); it permits the prediction of the most probable values of y.
Regression coefficient	The regression coefficient is the slope of the straight line that most closely relates two correlated variables.
Elements	A set is a collection of objects considered as a whole. The objects of a set are called elements or members. The elements of a set can be anything: numbers, people, letters of the alphabet, other sets, and so on.
Probability	Probability is the ratio of the number of favorable outcomes to the number of possible outcomes.
Multicolline-rity	Multicollinearity refers to linear inter-correlation among variables. Simply put, if nominally "different" measures actually quantify the same phenomenon to a significant degree -- i.e., wherein the variables are accorded different names and perhaps employ different numeric measurement scales but correlate highly with each other -- they are redundant.
Statistical significance	A result is has statistical significance if it is unlikely to have occurred by chance, given that a presumed null hypothesis is true.
Bias	A bias is a prejudice in a general or specific sense, usually in the sense for having a predilection to one particular point of view or ideological perspective. However, one is generally only said to be biased if one's powers of judgement are influenced by the biases one holds, to the extent that one's views could not be taken as being neutral or objective, but instead as subjective.
Factorial	The factorial of a natural number n is the product of all positive integers less than and equal to n. This is written as n! and pronounced "n factorial". The notation n! was introduced by Christian Kramp in 1808.
Population	A population is a set of entities concerning which statistical inferences are to be drawn, often based on a random sample taken from the population.
Sets	Sets are collections of objects considered as a whole. The objects of sets are called elements or members. The elements of a set can be anything: numbers, people, letters of the alphabet, other sets, and so on. Sets are conventionally denoted with capital letters, A, B, C, etc. Two sets A and B are said to be equal, written A = B, if they have the same members.
Mode	The mode is the value that has the largest number of observations, namely the most frequent value within a particular set of values.
Upper limit	The upper limit in a distribution is the greatest score or value in the distribution.

Go to **Cram101.com** for the Practice Tests for this Chapter.

Replication	Replication is repeating the creation of a phenomenon, so that the variability associated with the phenomenon can be estimated.
Interaction	Interaction is a kind of action which occurs as two or more objects have an effect upon one another. The idea of a two-way effect is essential in the concept of interaction instead of a one-way causal effect.
Confound	A confound is a "hidden" variable in a statistical or research model that affects the variables in question but is not known or acknowledged, and thus (potentially) distorts the resulting data. This hidden third variable causes the two measured variables to falsely appear to be in a causal relation.
Random sampling	In random sampling every combination of items from the frame, or stratum, has a known probability of occurring, but these probabilities are not necessarily equal. With any form of sampling there is a risk that the sample may not adequately represent the population but with random sampling there is a large body of statistical theory which quantifies the risk and thus enables an appropriate sample size to be chosen.
Skewness	Skewness is a measure of the asymmetry of the distribution of a real-valued random variable. Skewness, the third standardized moment, is written as $\gamma 1$ and defined as $\gamma 1 = \mu^3 / \sigma^3$ where μ^3 is the third moment about the mean and σ is the standard deviation.
Kurtosis	Kurtosis is a measure of the "peakedness" of the distribution of a real-valued random variable. Higher kurtosis means more of the variance is due to infrequent extreme deviations, as opposed to frequent modestly-sized deviations.
Outlier	In statistics, an outlier is a single observation "far away" from the rest of the data. In most samplings of data, some data points will be further away from their expected values than what is deemed reasonable. This can be due to systematic error or faults in the theory that generated the expected values.
Sample size	Sensitivity can be increased by using statistical controls, by increasing the reliability of measures (as in psychometric reliability), and by increasing the size of the sample. Increasing sample size is the most commonly used method for increasing statistical power.
Deviation	A deviation is the difference between an observed value and the expected value of a variable or function.
Standard deviation	The standard deviation is the most commonly used measure of statistical dispersion. Simply put, it measures how spread out the values in a data set are. The standard deviation is defined as the square root of the variance.
Power	The power of a statistical test is the probability that the test will reject a false null hypothesis, or in other words that it will not make a Type II error. The higher the power, the greater the chance of obtaining a statistically significant result when the null hypothesis is false.
Subset	A is a subset of a set B, if A is "contained" inside B. The relationship of one set being a subset of another is called inclusion. Every set is a subset of itself.
Internal consistency	Internal consistency focuses on the degree to which the individual items are correlated with each other and is thus often called homogeneity. Several statistics fall within this category. The best known are Cronbach's alpha, the Kuder-Richardson Formula 20 (KR-20) and the Kuder-Richardson Formula 21 (KR-21).
Significance level	The significance level of a test is the maximum probability of accidentally rejecting a true null hypothesis (a decision known as a Type I error). The significance of a result is also called its p-value; the smaller the p-value, the more significant the result is said to be.
Critical value	A critical value is the value corresponding to a given significance level. This cutoff value

Go to **Cram101.com** for the Practice Tests for this Chapter.
And, **NEVER** highlight a book again!

determines the boundary between those samples resulting in a test statistic that leads to rejecting the null hypothesis and those lead to a decision not to reject the null hypothesis.

Negative relationship

Negative relationship occurs when a change in one variable results in or is correlated with a change in another variable in the opposite direction. If the first increases, the second decreases, and so on.

Coefficient of determination

The coefficient of determination R^2 is the proportion of a sample variance of a response variable that is "explained" by the predictor variables when a linear regression is done.

Effect size

An effect size describes how large the relationship is between two variables. This information is important in scientific research. Often it is useful to know not only whether an experiment had an effect, but also the size of any effects.

Residual

Error is a misnomer; an error is the amount by which an observation differs from its expected value; the latter being based on the whole population from which the statistical unit was chosen randomly. A residual, on the other hand, is an observable estimate of the unobservable error.

Discriminant analysis

Linear discriminant analysis and the related Fisher's linear discriminant are used in machine learning to find the linear combination of features which best separate two or more classes of object or event.

Discriminant

A discriminant is an expression that discriminates qualities of algebraic structures. The concept applies to polynomials, conic sections, quadratic forms, and algebraic number fields.

Information theory

Information theory is a field of mathematics concerning the storage and transmission of data and includes the fundamental concepts of source coding and channel coding.

Lower limit

The lower limit in a distribution is the smallest value.

Analysis of covariance

Analysis of covariance is an old-fashioned name for a linear regression model with one continuous explanatory variable and one or more factors. The name exists for historical reasons, but there is no particular reason to distinguish the method from the general purpose linear model.

Goodness of fit

Goodness of fit means how well a statistical model fits a set of observations. Measures of goodness of fit typically summarize the discrepancy between observed values and the values expected under the model in question. Such measures can be used in statistical hypothesis testing, e.g. to test for normality of residuals, to test whether two samples are drawn from identical distributions.

Hypothesis

A hypothesis is a proposed explanation for a phenomenon. A scientific hypothesis must be testable and based on previous observations or extensions of scientific theories.

Hypothesis testing

Hypothesis testing is an algorithm to state the alternative which minimizes certain risks.

Test Statistic

A test statistic is a summary value (often a summary statistic) of a data set that is compared with a statistical distribution to determine whether the data set differs from that expected under a null hypothesis.

Multivariate	In statistics, in multivariate data, each data point has more than one scalar component, and often one is concerned with correlations between the components.
Multivariate analysis	Multivariate analysis in statistics describes a collection of procedures which involve observation and analysis of more than one statistical variable at a time.
Complement	Generally a complement of X is something that together with X makes a complete whole; that supplies what X lacks.
Confidence Interval	A confidence interval is an interval between two numbers, where there is a certain specified level of confidence that a population parameter lies.
Power	The power of a statistical test is the probability that the test will reject a false null hypothesis, or in other words that it will not make a Type II error. The higher the power, the greater the chance of obtaining a statistically significant result when the null hypothesis is false.
Parameter	A parameter is a characteristic of a population.
Association	In statistics, an association comes from two variables that are related. Many people confuse association with cauzation. Association does not imply cauzation.
Correlation	Correlation indicates the strength and direction of a linear relationship between two random variables. In general statistical usage, correlation refers to the departure of two variables from independence.
Elements	A set is a collection of objects considered as a whole. The objects of a set are called elements or members. The elements of a set can be anything: numbers, people, letters of the alphabet, other sets, and so on.
Bias	A bias is a prejudice in a general or specific sense, usually in the sense for having a predilection to one particular point of view or ideological perspective. However, one is generally only said to be biased if one's powers of judgement are influenced by the biases one holds, to the extent that one's views could not be taken as being neutral or objective, but instead as subjective.
Statistic	A statistic is a characteristic of a sample drawn from a population.
Decision tree	In decision theory (for example risk management), a decision tree is a graph of decisions and their possible consequences, (including resource costs and risks) used to create a plan to reach a goal.
Variable	A variable is a symbol denoting a quantity or symbolic representation. In mathematics, a variable often represents an unknown quantity; in computer science, it represents a place where a quantity can be stored.
Binary variable	A binary variable can only take on one of two values.
Mode	The mode is the value that has the largest number of observations, namely the most frequent value within a particular set of values.
Sample	A sample is that part of a population which is actually observed. In normal scientific practice, we demand that it is selected in such a way as to avoid presenting a biased view of the population.
Sampling distribution	A sampling distribution is the probability distribution, under repeated sampling of the population, of a given statistic (a numerical quantity calculated from the data values in a sample).
Distribution	A distribution is a list of the values that a variable takes in a sample. It is usually a list, ordered by quantity.

Go to **Cram101.com** for the Practice Tests for this Chapter.

Sampling	Sampling is that part of statistical practice concerned with the selection of individual observations intended to yield some knowledge about a population of concern, especially for the purposes of statistical inference.
Range	In descriptive statistics, the range is the length of the smallest interval which contains all the data. It is calculated by subtracting the smallest observations from the greatest and provides an indication of statistical dispersion.
Sample size	Sensitivity can be increased by using statistical controls, by increasing the reliability of measures (as in psychometric reliability), and by increasing the size of the sample. Increasing sample size is the most commonly used method for increasing statistical power.
Statistical Inference	Statistical inference is inference about a population from a random sample drawn from it or, more generally, about a random process from its observed behavior during a finite period of time. It includes: point estimation, interval estimation, hypothesis testing (or statistical significance testing) prediction
Probability distribution	Every random variable gives rise to a probability distribution, containing the most important information about the variable. If X is a random variable, the corresponding probability distribution assigns to the interval (a, b) the probability $Pr(a \leq X \leq b)$, i.e. the probability that the variable X will take a value in the interval (a, b).
Combinations	Combinations are un-ordered collections of unique elements. The order of the elements is not important
Probability	Probability is the ratio of the number of favorable outcomes to the number of possible outcomes.
Generalization	Concept A is a (strict) generalization of concept B if and only if: every instance of concept B is also an instance of concept A; and there are instances of concept A which are not instances of concept B.
Discriminant analysis	Linear discriminant analysis and the related Fisher's linear discriminant are used in machine learning to find the linear combination of features which best separate two or more classes of object or event.
Multiple regression	A multiple regression is a linear regression with more than one covariate (predictor variable). It can be viewed as a simple case of canonical correlation. An equation used to predict a dependent variable, y from two independents, u and v is: $y = \beta_0 + \beta_1 u + \beta_2 v + \beta_3 u2 + \beta_4 uv + \beta_5 v2$
Discriminant	A discriminant is an expression that discriminates qualities of algebraic structures. The concept applies to polynomials, conic sections, quadratic forms, and algebraic number fields.
Mean	For a real-valued random variable X, the mean is the expectation of X. If the expectation does not exist, then the random variable has no mean. For a data set, the mean is just the sum of all the observations divided by the number of observations.
Measurement	Measurement generally refers to the process of estimating or determining the ratio of a magnitude of a quantitative property or relation to a unit of the same type of quantitative property or relation.
Generalizability	Generalizability describes whether the results of individual studies and investigation samples can be applied to other studies and investigation samples.
Population	A population is a set of entities concerning which statistical inferences are to be drawn, often based on a random sample taken from the population.
Dependent variable	A variable that changes as a function of a change to another variable is called a dependent variable.

Go to **Cram101.com** for the Practice Tests for this Chapter.

Independent variable	An independent variable is presumed to cause or determine a dependent variable. It can be changed as required and its values do not represent a problem requiring explanation in an analysis, but are taken simply as given.
Mutually exclusive	In probability theory, events E_1, E_2, ..., E_n are said to be mutually exclusive if the occurrence of any one them automatically implies the non-occurrence of the remaining $n - 1$ events. In other words, two mutually exclusive events cannot both occur.
Sets	Sets are collections of objects considered as a whole. The objects of sets are called elements or members. The elements of a set can be anything: numbers, people, letters of the alphabet, other sets, and so on. Sets are conventionally denoted with capital letters, A, B, C, etc. Two sets A and B are said to be equal, written A = B, if they have the same members.
Interaction	Interaction is a kind of action which occurs as two or more objects have an effect upon one another. The idea of a two-way effect is essential in the concept of interaction instead of a one-way causal effect.
Construct	A construct is a mathematical or conceptual model.
Time series	In statistics and signal processing, a time series is a sequence of data points, measured typically at successive times, spaced apart at uniform time intervals. Time series analysis comprises methods that attempt to understand such time series, often either to understand the underlying theory of the data points, or to make forecasts.
Dummy variable	A dummy variable is a notation for a place or places in an expression, into which some definite substitution may take place, or with respect to which some operation (summation or quantification, to give two examples) may take place. The idea is related to, but somewhat deeper and more complex than, that of a placeholder (a symbol that will later be replaced by some literal string), or a wildcard character that stands for an unspecified symbol.
Residual	Error is a misnomer; an error is the amount by which an observation differs from its expected value; the latter being based on the whole population from which the statistical unit was chosen randomly. A residual, on the other hand, is an observable estimate of the unobservable error.
Skewness	Skewness is a measure of the asymmetry of the distribution of a real-valued random variable. Skewness, the third standardized moment, is written as $\gamma 1$ and defined as $\gamma 1 = \mu^3 / \sigma^3$ where μ^3 is the third moment about the mean and σ is the standard deviation.
Outlier	In statistics, an outlier is a single observation "far away" from the rest of the data. In most samplings of data, some data points will be further away from their expected values than what is deemed reasonable. This can be due to systematic error or faults in the theory that generated the expected values.
Regression equation	The regression equation represents the relation between selected values of one variable (x) and observed values of the other (y); it permits the prediction of the most probable values of y.
Interval estimate	An Interval estimate is a range of values estimated to include the parameter.
Normal distribution	The normal distribution is an extremely important probability distribution in many fields. It is a family of distributions of the same general form, differing in their location and scale parameters: the mean and standard deviation. The standard normal distribution is the normal distribution with a mean of zero and a standard deviation of one
Sample Mean	The arithmetic mean of a set of numbers is the sum of all the members of the set divided by the number of items in the set. If the set is a statistical population, then we speak of the population mean; if of a sampling of a population, it is a sample mean.

Go to **Cram101.com** for the Practice Tests for this Chapter.

Deviation	A deviation is the difference between an observed value and the expected value of a variable or function.
Standard deviation	The standard deviation is the most commonly used measure of statistical dispersion. Simply put, it measures how spread out the values in a data set are. The standard deviation is defined as the square root of the variance.
Standard error	The standard error of a measurement, value or quantity is the standard deviation of the process by which it was generated, after adjusting for sample size.
Central limit theorem	The Central Limit Theorem states that if the sum of the variables has a finite variance, then it will be approximately normally distributed. Since many real processes yield distributions with finite variance, this explains the ubiquity of the normal distribution.
Theorem	A theorem is a proposition that has been or is to be proved on the basis of explicit assumptions.
Histogram	A histogram is a graphical display of tabulated frequencies. That is, a histogram is the graphical version of a table which shows what proportion of cases fall into each of several or many specified categories. The categories are usually specified as nonoverlapping intervals of some variable. The categories (bars) must be adjacent.
Sampling without replacement	Sampling without replacement means that in each successive trial of an experiment or process, the total number of possible outcomes or the mix of possible outcomes is changed by sampling. The probability of future events is thus changed.
Subset	A is a subset of a set B, if A is "contained" inside B. The relationship of one set being a subset of another is called inclusion. Every set is a subset of itself.
Estimator	An estimator is a function of the known sample data that is used to estimate an unknown population parameter; an estimate is the result from the actual application of the function to a particular set of data. Many different estimators are possible for any given parameter.
Statistical significance	A result is has statistical significance if it is unlikely to have occurred by chance, given that a presumed null hypothesis is true.
Multicollinerity	Multicollinearity refers to linear inter-correlation among variables. Simply put, if nominally "different" measures actually quantify the same phenomenon to a significant degree -- i.e., wherein the variables are accorded different names and perhaps employ different numeric measurement scales but correlate highly with each other -- they are redundant.
Bootstrapping	Bootstrapping is a statistical method for estimating the sampling distribution of an estimator by sampling with replacement from the original sample, most often with the purpose of deriving robust estimates of standard errors and confidence intervals of a population parameter like a mean, median, proportion, odds ratio, correlation coefficient or regression coefficient.

104

Go to **Cram101.com** for the Practice Tests for this Chapter.

Lightning Source UK Ltd.
Milton Keynes UK
UKOW05f2253270917

310018UK00002B/22/P

9 781428 813755